Nonlinear
Programming

McGraw-Hill Series in Systems Science

Editorial Consultants

A. V. Balakrishnam
George Dantzig
Lotfi Zadeh

Olvi L. Mangasarian

Professor of Computer Sciences
University of Wisconsin

Nonlinear Programming

McGraw-Hill Book Company

New York St. Louis San Francisco London
Sydney Toronto Mexico Panama

Nonlinear Programming

Library of Congress Catalog Card Number 69–17186

ISBN 07-039885-2

56789 KPKP 798765

To

Josephine Mangasarian, my mother,
and to
Claire

Preface

This book is based on a course in nonlinear programming given in the Electrical Engineering and Computer Sciences Department and the Industrial Engineering and Operations Research Department of the University of California at Berkeley and in the Computer Sciences Department of the University of Wisconsin at Madison. The intent of the book is to cover the fundamental theory underlying nonlinear programming for the applied mathematician. The entire book could be used as a text for a one-semester course, or the first eight chapters for a one-quarter course. The course level would probably be advanced undergraduate or first-year graduate. The only prerequisite would be a good course in advanced calculus or real analysis. (Linear programming is *not* a prerequisite.) All the results needed in the book are given in the Appendixes.

I am indebted to J. Ben Rosen who first introduced me to the fascinating subject of nonlinear programming, to Lotfi A. Zadeh who originally suggested the writing of such a book, to Jean-Paul Jacob, Phillippe Rossi, and James W. Daniel who read the manuscript carefully and made numerous improvements, and to all my students whose questions and observations resulted in many changes.

Olvi L. Mangasarian

To the Reader

The following system of numbering and cross-referencing is used in this book. At the top of each page in the outer margin appear chapter and section numbers in boldface type; for example, **3.2** at the top of a page means that the discussion on that page is part of Chapter *3*, Section *2*. In addition, each item on the page (Definition, Theorem, Example, Comment, Remark, etc.) is given a number that appears in the left-hand margin; such items are numbered consecutively within each section. Item numbers and all cross-references in the text are in italic type. Cross-references are of the form "by Definition *5.4.3*"; this means "by the definition which is item *3* of Section *4* in Chapter *5*." Since the four appendixes are labeled A, B, C, and D, the reference "*C.1.3*" is to "item *3* in Section *1*, Appendix C." When we refer in a section to an item within the same section, only the item number is given; thus "substituting in *7*" means "substituting in Equation *7* of this section."

Contents

Nonlinear
Programming

Chapter One

The Nonlinear Programming Problem, Preliminary Concepts, and Notation

1. The nonlinear programming problem†

The nonlinear programming problem that will concern us has three fundamental ingredients: a finite number of real variables, a finite number of constraints which the variables must satisfy, and a function of the variables which must be minimized (or maximized). Mathematically speaking we can state the problem as follows: Find specific values $(\bar{x}_1, \ldots, \bar{x}_n)$, if they exist, of the variables $(x_1, \ldots, x_n)$ that will satisfy the *inequality constraints*

1
$$g_i(x_1, \ldots, x_n) \leqq 0$$
$$i = 1, \ldots, m$$

the *equality constraints*

2
$$h_j(x_1, \ldots, x_n) = 0$$
$$j = 1, \ldots, k$$

and minimize (or maximize) the *objective function*

3
$$\theta(x_1, \ldots, x_n)$$

over all values of $x_1, \ldots, x_n$ satisfying *1* and *2*. Here, g_i, h_j, and θ are numerical functions‡ of the variables $x_1, \ldots, x_n$, which are defined for all finite values of

† In order to introduce the problem in the first section of the book, some undefined terms (function, real variable, constraints, etc.) must be interpreted intuitively for the time being. The problem will be stated rigorously at the end of this chapter (see *1.6.9* to *1.6.12*).

‡ The concept of a numerical function will be defined precisely in Sec. *1.5*. For the present by a numerical function of $x_1, \ldots, x_n$ we mean a correspondence which assigns a single real number for each n-tuple of real values that the variables $x_1, \ldots, x_n$ assume.

the variables. The fundamental difference between this problem and that of the classical constrained minimization problem of the ordinary calculus [Courant 47, Fleming 65]† is the presence of the inequalities 1. As such, inequalities will play a crucial role in nonlinear programming and will be studied in some detail.

As an example of the above problem consider the case shown in Fig. *1.1.1*. Here we have $n = 2$ (two variables x_1, x_2), $m = 3$ (three inequality constraints), and $k = 1$ (one equality constraint). Each curve in Fig. *1.1.1* is obtained by setting some numerical function equal to a real number such as $\theta(x_1, x_2) = 5$ or $g_2(x_1, x_2) = 0$. The little arrows on the

† This refers to the works by Courant, written in 1947, and by Fleming, written in 1965, as listed in the Bibliography at the back of the book. This system of references will be used throughout the book with one exception: [Gordan 73] refers to Gordan's paper written in 1873.

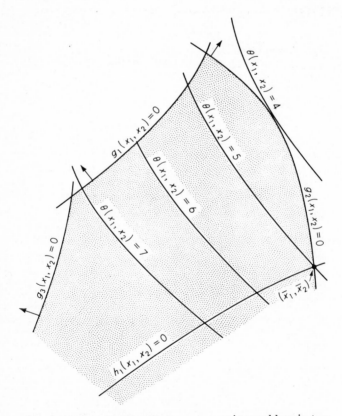

Fig. 1.1.1 A typical nonlinear programming problem in two
variables (x_1, x_2).

curves $g_i(x_1,x_2) = 0$ indicate the side in the direction of which g_i increases, and hence all (x_1,x_2) must lie on the opposite side of these curves if they are to satisfy 1. All such (x_1,x_2) lie in the shaded area of Fig. *1.1.1*. To satisfy *2*, (x_1,x_2) must lie on the curve $h_1(x_1,x_2) = 0$. The solution to the problem is $(\bar{x}_1,\bar{x}_2)$. This is the point on the curve $h_1(x_1,x_2) = 0$ at which θ assumes its lowest value over the set of all (x_1,x_2) satisfying $g_i(x_1,x_2) \leq 0$, $i = 1, 2, 3$. In more complicated situations where n, m, and k may be large, it will not be easy to solve the above problem. We shall then be concerned with obtaining necessary and/or sufficient conditions that a point $(\bar{x}_1, \ldots ,\bar{x}_n)$ must satisfy in order for it to solve the nonlinear programming problem *1* to *3*. These optimality conditions form the crux of nonlinear programming.

In dealing with problems of the above type we shall confine ourselves to minimization problems only. Maximization problems can be easily converted to minimization problems by employing the identity

4 $$\text{maximum } \theta(x_1, \ldots ,x_n) = -\text{minimum } [-\theta(x_1, \ldots ,x_n)]$$

5   ### Problem

Solve graphically as indicated in Fig. *1.1.1* the following nonlinear programming problem:

minimize $(-x_1 - x_2)$

subject to

$2(x_1)^2 - x_2 \leq 0$

$(x_1)^2 + (x_2)^2 = 1$

2. Sets and symbols

We shall use some symbols and elementary concepts from set theory [Anderson-Hall 63, Hamilton-Landin 61, Berge 63]. In particular a *set* Γ is a collection of objects of any kind which are by definition *elements* or *points* of Γ. For example if we let R (the *reals* or the *real line*) denote the set of all real numbers, then 7 is an element or point of R. We use the symbol $\in$ to denote the fact that an element belongs to a set. For example we write $7 \in R$. For simplicity we also write sometimes $5,7 \in R$ instead of $5 \in R$ and $7 \in R$.

If Γ and Λ are two sets, we say that Γ is *contained* in Λ, Γ is *in* Λ, Γ is a *subset of* Λ, or Λ *contains* Γ, if each element of Γ is also an element of Λ, and we write $\Gamma \subset \Lambda$ or $\Lambda \supset \Gamma$. If $\Gamma \subset \Lambda$ and $\Lambda \subset \Gamma$ we write $\Gamma = \Lambda$. A slash across a symbol denotes its negation. Thus $x \notin \Gamma$ and $\Gamma \not\subset \Lambda$ denote respectively that x is not an element of Γ and that Γ is not a subset of Λ. The *empty set* is the set which contains no elements and is denoted by $\emptyset$. We denote a set sometimes by $\{x,y,z\}$, if the set is formed by the elements x, y, z. Sometimes a set is characterized by a property that its elements must have, in which case we write

$\{x \mid x$ satisfying property $P\}$

For example the set of all nonnegative real numbers can be written as

$\{x \mid x \in R, x \geqq 0\}$

The set of elements belonging to either of two sets Γ or Λ is called the *union* of the sets Γ and Λ and is denoted by $\Gamma \cup \Lambda$. We have then

$$\Gamma \cup \Lambda = \{x \mid x \in \Gamma \text{ or } x \in \Lambda\}$$

The set of elements belonging to at least one of the sets of the (finite or infinite) family of sets $(\Gamma_i)_{i \in I}$ is called the *union of the family* and is denoted by $\underset{i \in I}{\cup} \Gamma_i$. Then

$$\underset{i \in I}{\cup} \Gamma_i = \{x \mid x \in \Gamma_i \text{ for some } i \in I\}$$

The set of elements belonging to both sets Γ and Λ is called the *intersection* of the sets Γ and Λ and is denoted by $\Gamma \cap \Lambda$. We then have

$$\Gamma \cap \Lambda = \{x \mid x \in \Gamma \text{ and } x \in \Lambda\}$$

The set of elements belonging to all the sets of the (finite or infinite) family of sets $(\Gamma_i)_{i \in I}$, is called the *intersection of the family* and is denoted by $\underset{i \in I}{\cap} \Gamma_i$. Then

$$\underset{i \in I}{\cap} \Gamma_i = \{x \mid x \in \Gamma_i \text{ for each } i \in I\}$$

Two sets Γ and Λ are *disjoint* if they do not intersect, that is, if $\Gamma \cap \Lambda = \emptyset$.

The *difference* of the sets Λ and Γ is the set of those elements of Λ not contained in Γ and is denoted by $\Lambda \sim \Gamma$. We have then

$$\Lambda \sim \Gamma = \{x \mid x \in \Lambda, x \notin \Gamma\}$$

In the above it is not assumed in general that $\Gamma \subset \Lambda$. If however $\Gamma \subset \Lambda$, then $\Lambda \sim \Gamma$ is called the *complement of Γ relative to Λ*.

The *product* of two sets Γ and Λ, denoted by $\Gamma \times \Lambda$, is defined as the set of ordered pairs (x,y) of which $x \in \Gamma$ and $y \in \Lambda$. We have then

$$\Gamma \times \Lambda = \{(x,y) \mid x \in \Gamma, y \in \Lambda\}$$

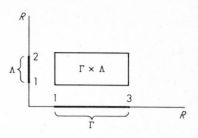

Fig. 1.2.1 The product $\Gamma \times \Lambda$ of the sets Γ
and Λ.

The *product of n sets* $\Gamma_1, \ldots, \Gamma_n$, denoted by $\Gamma_1 \times \Gamma_2 \times \cdots \times \Gamma_n$, is defined as the set of ordered n-tuples $(x_1, \ldots, x_n)$ of which $x_1 \in \Gamma_1$, $\ldots, x_n \in \Gamma_n$. We have then

$$\Gamma_1 \times \Gamma_2 \times \cdots \times \Gamma_n = \{(x_1, \ldots, x_n) \mid x_1 \in \Gamma_1, \ldots, x_n \in \Gamma_n\}$$

If $\Gamma_1 = \Gamma_2 = \cdots = \Gamma_n = \Gamma$, then we write $\Gamma^n = \Gamma \times \Gamma \times \cdots \times \Gamma$. If we let

$$\Gamma = \{x \mid x \in R, 1 \leqq x \leqq 3\} \qquad \Lambda = \{x \mid x \in R, 1 \leqq x \leqq 2\}$$

then

$$\Gamma \times \Lambda = \{(x,y) \mid x \in R, y \in R, 1 \leqq x \leqq 3, 1 \leqq y \leqq 2\}$$

Figure *1.2.1* depicts the set $\Gamma \times \Lambda$. The set $R^2 = R \times R$, which can be represented by points on a plane, is called the *Euclidean plane*.
The following symbols will also be used:

$(\forall x)$ reads *for each x*
$(\exists x)$ reads *there exists an x such that*
$\Rightarrow$ reads *implies*
$\Leftarrow$ reads *is implied by*
$\Leftrightarrow$ reads *is equivalent to*
(A slash [/] across any one of the last three symbols denotes their negation.)

For example the statement "for each x there exists a y such that $\theta(x,y) = 1$" can be written as

$$(\forall x)(\exists y): \theta(x,y) = 1$$

The negation of the above statement can be automatically written as

$$(\exists x)(\forall y): \theta(x,y) \neq 1$$

Frequently we shall refer to a certain relationship such as an equation or an inequality by a number or Roman numeral such as I or II.

The notation $I \Rightarrow II$ means relationship I implies relationship II. An overbar on I or II ($\bar{I}$ or $\overline{II}$) denotes the negation of the relationship referred to by that numeral. Obviously then the statement that $I \Rightarrow II$ is logically equivalent to $\bar{I} \Leftarrow \overline{II}$. Thus

$$\langle I \Rightarrow II \rangle \Leftrightarrow \langle \bar{I} \Leftarrow \overline{II} \rangle$$

3. Vectors

1
n-vector

An *n-vector* or *n-dimensional vector* x, for any positive integer n, is an n-tuple $(x_1, \ldots, x_n)$ of real numbers. The real number x_i is referred to as the *ith component* or *element* of the vector x.

2
R^n

The *n-dimensional (real) Euclidean space* R^n, for any positive integer n, is the set of all n-vectors.

The notation $x \in R^n$ means that x is an element of R^n, and hence, x is an n-vector. Frequently we shall also refer to x as a *point* in R^n. R^1, or simply R, is then the Euclidean line (the set of all real numbers), R^2 is the Euclidean plane (the set of all ordered pairs of real numbers), and $R^n = R \times R \times \cdots \times R$ (n times).

3
Vector addition and multiplication by a real number

Let $x, y \in R^n$ and $\alpha \in R$. The *sum* $x + y$ is defined by

$$x + y = (x_1 + y_1, \ldots, x_n + y_n)$$

and the *multiplication by a real number* αx is defined by

$$\alpha x = (\alpha x_1, \ldots, \alpha x_n)$$

4
Linear dependence and independence

The vectors $x^1, \ldots, x^m \in R^n$ are said to be *linearly independent* if

$$\left. \begin{array}{c} \lambda^1 x^1 + \cdots + \lambda^m x^m = 0 \\ \lambda^1, \ldots, \lambda^m \in R \end{array} \right\} \Rightarrow \lambda^1 = \lambda^2 = \cdots = \lambda^m = 0$$

otherwise they are *linearly dependent*. (Here and elsewhere 0 denotes the real number zero or a vector each element of which is zero.)

5
Linear combination

The vector $x \in R^n$ is a *linear combination* of $x^1, \ldots, x^m \in R^n$ if

$$x = \lambda^1 x^1 + \cdots + \lambda^m x^m \qquad \text{for some } \lambda^1, \ldots, \lambda^m \in R$$

and it is a *nonnegative linear combination* of $x^1, \ldots, x^m$ if in addition to the above equality $\lambda^1, \ldots, \lambda^m \geqq 0$. The numbers $\lambda^1, \ldots, \lambda^m$ are called *weights*.

The above concepts involving vector addition and multiplication by a scalar define the vector space structure of R^n. They are not enough however to define the concept of distance. For that purpose we introduce the scalar product of two vectors.

6   ### Scalar product

The *scalar product* xy of two vectors $x, y \in R^n$ is defined by

$$xy = x_1 y_1 + \cdots + x_n y_n$$

7   ### Norm of a vector

The *norm* $\|x\|$ *of a vector* $x \in R^n$ is defined by

$$\|x\| = +(xx)^{\frac{1}{2}} = +[(x_1)^2 + \cdots + (x_n)^2]^{\frac{1}{2}}$$

8   ### Cauchy-Schwarz inequality

Let $x, y \in R^n$. *Then*

$$|xy| \leqq \|x\| \cdot \|y\|$$

where $|xy|$ *is the absolute value of the real number* xy.

PROOF   Let $x, y \in R^n$ be fixed. For any $\alpha \in R$

$$(\alpha x + y)(\alpha x + y) = xx(\alpha)^2 + 2xy\alpha + yy \geqq 0$$

Hence the roots of the quadratic equation in α

$$xx(\alpha)^2 + 2xy\alpha + yy = 0$$

cannot be distinct real numbers, and so

$$(xy)^2 \leqq (xx)(yy)$$

which implies the Cauchy-Schwarz inequality.   ∎

9   ### Distance between two points

Let $x, y \in R^n$. The nonnegative number $\delta(x,y) = \|x - y\|$ is called the *distance between the two points* x *and* y in R^n.

10   ### Problem

Establish the fact that R^n is a *metric space* by showing that

$\delta(x,y)$ satisfies the following conditions

$\delta(x,y) \geqq 0$

$\delta(x,y) = 0 \Leftrightarrow x = y$

$\delta(x,y) = \delta(y,x)$

$\delta(x,z) \leqq \delta(x,y) + \delta(y,z)$ (triangle inequality)

(Hint: Use the Cauchy-Schwarz inequality to establish the triangle inequality.)

11   **Angle between two vectors**

Let x and y be two nonzero vectors in R^n: The *angle* ψ *between* x *and* y is defined by the formula

$$\cos \psi = \frac{xy}{\|x\| \cdot \|y\|} \qquad 0 \leq \psi \leq \pi$$

This definition of angle agrees for $n = 2,3$ with the one in analytic geometry. The nonzero vectors x and y are *orthogonal* if $xy = 0$ ($\psi = \pi/2$); form an *acute angle* with each other if $xy \geqq 0$ ($0 \leqq \psi \leqq \pi/2$), a *strict acute angle* if $xy > 0$ ($0 \leqq \psi < \pi/2$), an *obtuse angle* if $xy \leqq 0$ ($\pi/2 \leqq \psi \leqq \pi$), and a *strict obtuse angle* if $xy < 0$ ($\pi/2 < \psi \leqq \pi$).

4. Matrices

Although our main concern is nonlinear problems, linear systems of the following type will be encountered very frequently:

$A_{11}x_1 + \cdots + A_{1n}x_n = b_1$

1 $\cdots \cdots \cdots \cdots \cdots \cdots \cdots$

$A_{m1}x_1 + \cdots + A_{mn}x_n = b_m$

where A_{ij} and b_i, $i = 1, \ldots, m$, $j = 1, \ldots, n$, are given real numbers. We can abbreviate the above system by using the concepts of the previous section. If we let A_i denote an n-vector whose n components are A_{ij}, $j = 1, \ldots, n$, and if we let $x \in R^n$, then the above system is equivalent to

2 $A_i x = b_i \qquad i = 1, \ldots, m$

In *2* we interpret $A_i x$ as the scalar product *1.3.6* of A_i and x. If we further let Ax denote an m-vector whose m components are $A_i x$, $i = 1, \ldots, m$, and b an m-vector whose m components are b_i, then the equivalent systems *1* and *2* can be further simplified to

3 $Ax = b$

In order to be consistent with ordinary matrix theory notation, we define the $m \times n$ *matrix* A as follows

4   $$A = \begin{pmatrix} A_{11} & \cdots & A_{1n} \\ \cdots & \cdots & \cdots \\ A_{m1} & \cdots & A_{mn} \end{pmatrix}$$

The *ith row* of the matrix A will be denoted by A_i and will be an n-vector. Hence

5   $A_i = (A_{i1}, A_{i2}, \ldots, A_{in}) \qquad i = 1, \ldots, m$

The *jth column* of the matrix A will be denoted by $A_{.j}$ and will be an m-vector.   Hence

6   $$A_{.j} = \begin{pmatrix} A_{1j} \\ A_{2j} \\ \cdot \\ \cdot \\ \cdot \\ A_{mj} \end{pmatrix}$$

The transpose of the matrix A is denoted by A' and is defined by

7   $$A' = \begin{pmatrix} A_{11} & \cdots & A_{m1} \\ \cdots & \cdots & \cdots \\ A_{1n} & \cdots & A_{mn} \end{pmatrix}$$

Obviously the ith row of A is equal to the ith column of A', and the jth column of A is equal to the jth row of A'.   Hence

8   $A_i = (A')_{.i} = A'_{.i}$

9   $A_{.j} = (A')_j = A'_j$

The last equalities of *8* and *9* are to be taken as the definitions of $A'_{.i}$ and A'_j respectively.   Since A_{ij} is the real number in the ith row of the jth column of A, then if we define A'_{ji} as the real number in the jth row of the ith column of A', we have

10   $A_{ij} = A'_{ji}$

The equivalent systems *1*, *2*, and *3* can be written still in another form as follows

11   $$\sum_{j=1}^{n} A_{.j} x_j = b$$

Here $A_{.j}$ and b are vectors in R^m and x_j are real numbers.   The representation 2 can be interpreted as a problem in R^n whereas 11 can be interpreted as a problem in R^m.   In 2 we are required to find an $x \in R^n$ that makes the appropriate scalar products b_i (or angles, see $1.3.11$) with the n-vectors A_i, $i = 1, \ldots, m$.   In 11, we are given the $n + 1$ vectors in R^m, $A_{.j}$, $j = 1, \ldots, n$ and b, and we are required to find n weights $x_1, \ldots, x_n$ such that b is a linear combination of the vectors $A_{.j}$. These two dual representations of the same linear system will be used in interpreting some of the important theorems of the alternative of the next chapter.

The $m \times n$ matrix A of 4 can generate another linear system yA, defined as follows

$$12 \qquad yA = A'y = \sum_{i=1}^{m} A'_{.i} y_i = \sum_{i=1}^{m} A_i y_i$$

where $y \in R^m$.   Hence, yA is an n-dimensional vector whose jth component is given by

$$13 \qquad (yA)_j = A_{.j} y \qquad j = 1, \ldots, n$$

In general we shall follow the convention of using upper case Latin letters to denote matrices.   If A is an $m \times n$ matrix, and if we let

$$14 \quad I \subset M = \{1, 2, \ldots, m\}$$

$$15 \quad J \subset N = \{1, 2, \ldots, n\}$$

then we define the following *submatrices of A* (which are matrices with rows and columns extracted respectively from the rows and columns of A)

$$16 \qquad A_I = \{A_i \mid i \in I\}$$

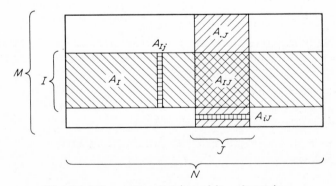

Fig. 1.4.1 An $m \times n$ matrix and its submatrices.

17 $\quad A_{.J} = \{A_{.j} \mid j \in J\}$

18 $\quad A_{Ij} = j$th column of A_I

19 $\quad A_{iJ} = i$th row of $A_{.J}$

20 $\quad A_{IJ} = (A_{Ij})_{j \in J} = (A_{iJ})_{i \in I}$

It follows then that

21 $\quad A_M = A_{.N} = A$

22 $\quad A_{IN} = A_I \qquad$ and $\qquad A_{MJ} = A_{.J}$

Figure *1.4.1* depicts some of the above submatrices of A.

23 $\qquad$ **Nonvacuous matrix**

$\qquad$ A matrix A is said to be *nonvacuous* if it contains at least one element A_{ij}. An $m \times n$ matrix A with $m \geq 1$ and $n \geq 1$ is nonvacuous even if all its elements $A_{ij} = 0$.

5. Mappings and functions

1 $\qquad$ **Mapping**

$\qquad$ Let X and Y be two sets. A *mapping* Γ *from X into Y* is a correspondence which associates with every element x of X a subset of Y. For each $x \in X$, the set $\Gamma(x)$ is called the *image* of x. The *domain* X^* of Γ is the subset of points of X for which the image $\Gamma(x)$ is nonempty, that is,

$$X^* = \{x \mid x \in X, \Gamma(x) \neq \emptyset\}$$

The *range* $\Gamma(X^*)$ of Γ is the union of the images of all the points of X^*, that is

$$\Gamma(X^*) = \bigcup_{x \in X^*} \Gamma(x)$$

EXAMPLE Let $X = Y = R$, $\Gamma(x) = \{y \mid \cos y = x\}$. Then $X^* = \{x \mid x \in R, -1 \leq x \leq 1\}$, $\Gamma(X^*) = Y = R$.

2 $\qquad$ **Function**

$\qquad$ A *function f* is a single-valued mapping from a set X into a set Y. That is for each $x \in X$, the image set $f(x)$ consists of a single element of Y. The domain of f is X, and we say that f is *defined on X*. The range of f is $f(X) = \bigcup_{x \in X} f(x)$. (For convenience we will write the image of a function not as a set but as the unique element of that set.)

3   Numerical function

A *numerical function* θ is a function from a set X into R. In other words a numerical function is a correspondence which associates a real number with each element x of X.

EXAMPLES If $X = R$, then θ is the familiar real single-valued function of a real variable, such as $\theta(x) = \sin x$. If X is the set of positive integers, then θ assigns a real number for each positive integer, for example $\theta(x) = 1/x!$. If $X = R^n$, then θ is the real single-valued function of n variables.

4   Vector function

An *m-dimensional vector function* f is a function from a set X into R^m. In other words a vector function is a correspondence which associates a vector from R^m with each element x of X. The m components of the vector $f(x)$ are denoted by $f_1(x)$, . . . , $f_m(x)$. Each f_i is a numerical function on X. A vector function f has a certain property (for example continuity) whenever each of its components f_i has that property.

EXAMPLE If $X = R^n$, then f associates a point of R^m with each point of R^n. The m components f_i, $i = 1$, . . . , m, of f are numerical functions on R^n.

5   Linear vector functions on R^n

An m-dimensional vector function f defined on R^n is said to be *linear* if

$$f(x) = Ax + b$$

where A is some fixed $m \times n$ matrix and b is some fixed vector in R^m.
It follows that if f is a linear function on R^n then

$$f(x^1 + x^2) = f(x^1) + f(x^2) - f(0) \qquad \text{for } x^1, x^2 \in R^n$$

$$f(\lambda x) = \lambda f(x) + (1 - \lambda)f(0) \qquad \text{for } \lambda \in R, \, x \in R^n$$

(Conversely, the last two relations could be used to define a linear vector function on R^n, from which it could be shown that $f(x) = Ax + b$ [Berge 63, p. 159].)
If $m = 1$ in the above, then we have a *numerical linear function* θ on R^n and

$$\theta(x) = cx + \gamma$$

where c is a fixed vector in R^n and γ is a fixed real number.

Inequalities or equalities involving linear vector functions (or linear numerical functions) will be naturally called *linear inequalities* or *equalities*.

6. Notation

1 Vectors and real numbers

In general we shall follow the convention that small Latin letters will denote vectors such as a, b, c, x, y, z, or vector functions such as f, g, h. Exceptions will be the letters i, j, k, m, n, and sometimes others, which will denote integers. Small Greek letters will denote a real number (a point in R) such as α, β, γ, ξ, η, ζ, or a numerical function such as θ, ϕ, ψ.

2 Subscripts

A small Latin letter with an integer subscript or a small Latin letter subscript will denote a component of a vector, in general, and on occasion will denote a vector. For example, if $x \in R^5$, then x_3 and x_i denote respectively the third and ith components of x. On the other hand we will have occasion to let $x_1 \in R^m$, $x_2 \in R^m$, etc., in which case this intent will be made explicit. Small Greek letters with integer or Latin subscripts will occasionally be used to denote real numbers such as λ_1, λ_i. If $x \in R^n$, $K \subset N = \{1, \ldots, n\}$, and K contains k elements each of which is distinct, then $x_{i \in K}$ is a vector in R^k with the components $\{x_i \mid i \in K\}$ and is denoted by x_K. Thus a small Latin letter with a capital Latin letter subscript denotes a vector in a space of smaller or equal dimension to that of the space of the unsubscripted vector.

3 Superscripts

A small Latin or Greek letter with a superscript or an elevated symbol will denote a fixed vector or real number, for example x^1, x^2, x^i, $\bar{x}$, $\hat{x}$, ξ^1, $\bar{\xi}$, etc. Exponentiation on the other hand will be distinguished by parentheses enclosing the quantity raised to a power, for example $(x)^2$.

4 Zero

The number 0 will denote either the real number zero or a vector in R^n all components of which are zero.

5 Matrices

Matrices will be denoted by capital Latin letters as described in detail in a previous section, Sec. *1.4*.

6   Sets

Sets will always be denoted by capital Greek or Latin letters such as Γ, Λ, Ω, R, I, X, Y. Capital letters with subscripts, such as Γ_1, Γ_2, Γ_i, and capital letters with elevated symbols, such as Γ^*, X^0 will also denote sets. (See also Sec. *1.2*.)

7   Ordering relations

The following convention for equalities and inequalities will be used. If $x, y \in R^n$, then

$$x = y \Leftrightarrow x_i = y_i \qquad i = 1, \ldots, n$$

$$x \geqq y \Leftrightarrow x_i \geqq y_i \qquad i = 1, \ldots, n$$

$$x \geq y \Leftrightarrow x \geqq y \qquad \text{and} \qquad x \neq y$$

$$x > y \Leftrightarrow x_i > y_i \qquad i = 1, \ldots, n$$

If $x \geqq 0$, x is said to be *nonnegative*, if $x \geq 0$ then x is said to be *semipositive*, and if $x > 0$ then x is said to be *positive*. The relations $=$, $\geqq$, $\geq$, $>$ defined above are called *ordering relations* (in R^n).

8   The nonlinear programming problem

By using the notation introduced above, the nonlinear programming problem *1.1.1* to *1.1.3* can be rewritten in a slightly more general form as follows. Let $X^0 \subset R^n$, let g, h, and θ be respectively an m-dimensional vector function, a k-dimensional vector function, and a numerical function, all defined on X^0. Then the problem becomes this: Find an $\bar{x}$, if such exists, such that

9 $\theta(\bar{x}) = \min\limits_{x \in X} \theta(x) \qquad \bar{x} \in X = \{x \mid x \in X^0, g(x) \leqq 0, h(x) = 0\}$

The set X is called the *feasible region*, $\bar{x}$ the *minimum solution*, and $\theta(\bar{x})$ the *minimum*. All points x in the feasible region X are referred to as *feasible points* or simply as *feasible*.

Another way of writing the same problem which is quite common in the literature is the following:

10 $\min\limits_{x \in X^0} \theta(x)$

subject to

11 $g(x) \leqq 0$

12 $h(x) = 0$

We favor the more precise and brief designation 9 of the problem instead of 10 to 12. Notice that if we let $X^0 = R^n$ in the above problem, then we obtain the nonlinear programming problem $1.1.1$ to $1.1.3$.

If $X^0 = R^n$ and θ, g, and h are all linear functions on R^n, then problem 9 becomes a *linear programming* problem: Find an $\bar{x}$, if such exists, such that

$$3 \qquad -b\bar{x} = \min_{x \in X} (-bx) \qquad \bar{x} \in X = \{x \mid x \in R^n,\ Ax \leqq c,\ Bx = d\}$$

where b, c, and d are given fixed vectors in R^n, R^m, and R^k respectively, and A and B are given fixed $m \times n$ and $k \times n$ matrices respectively. There exists a vast literature on the subject of linear programming [Dantzig 63, Gass 64, Hadley 62, Simmonard 66]. It should be remarked that problem 13 is equivalent to finding an $\bar{x}$ such that

$$4 \qquad b\bar{x} = \max_{x \in X} bx \qquad \bar{x} \in X = \{x \mid x \in R^n,\ Ax \leqq c,\ Bx = d\}$$

When B and d are absent from this formulation, 14 becomes the standard dual form of the linear programming problem [Simmonard 66, p. 95].

Chapter Two

Linear Inequalities and Theorems of the Alternative

1. Introduction

It was mentioned in Chap. 1 that the presence of inequality constraints in a minimization problem constitutes the distinguishing feature between the minimization problems of the classical calculus and those of nonlinear programming. Although our main interest lies in nonlinear problems, and hence in nonlinear inequalities, linearization (that is, approximating nonlinear constraints by linear ones) will be frequently resorted to. This will lead us to linear inequalities. It is the purpose of this chapter to establish some fundamental theorems for linear inequalities which will be used throughout this work. (Needless to say, these fundamental theorems also play a crucial role in linear programming. See for example [Gale 60, chap. 2].)

The type of theorem that will concern us in this chapter will involve two systems of linear inequalities and/or equalities, say systems I and II. A typical theorem of the alternative asserts that either system I has a solution, or that system II has a solution, but never both. The most famous theorem of this type is perhaps Farkas' theorem [Farkas 02, Tucker 56, Gale 60].

Farkas' theorem

For each fixed $p \times n$ matrix A and each fixed vector b in R^n, either

I $\qquad Ax \leqq 0 \qquad bx > 0$
$\qquad\qquad$ *has a solution $x \in R^n$*

or

1

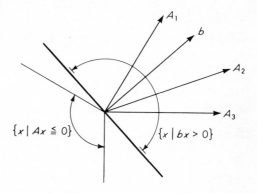

Fig. 2.1.1 Geometric interpretation
of Farkas' theorem: II'
has solution, I' has no
solution.

II $A'y = b$ $y \geqq 0$ *has a solution* $y \in R^p$

but never both.

   We shall postpone a proof of Farkas' theorem until after we have
given a geometric interpretation of it and applied it to get the necessary
optimality conditions of linear programming.

   To give a geometric interpretation of Farkas' theorem, we rewrite
I and II as follows

I' $A_j x \leqq 0, j = 1, \ldots , p, bx > 0$

II' $\sum_{j=1}^{p} A'_{.j} y_j = \sum_{j=1}^{p} A_j y_j = b, y_j \geqq 0, j = 1, \ldots , p$

where $A'_{.j}$ denotes the jth column of A' and A_j the jth row of A (see
Sec. *1.4*).   System II' requires that the vector b be a nonnegative linear
combination of the vectors A_1 to A_p.   System I' requires that we find a
vector $x \in R^n$ that makes an obtuse angle ($\geqq \pi/2$) with the vectors A_1
to A_p and a strictly acute angle ($< \pi/2$) with b (see *1.3.11*).   Figure *2.1.1*

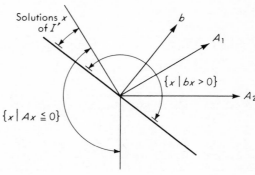

Fig. 2.1.2 Geometric interpretation of Farkas'
theorem: I' has so-
lution, II' has no
solution.

shows a simple case with $n = 2$, $p = 3$, in which II′ has a solution, and hence by Farkas' theorem I′ cannot have a solution. Figure 2.1.2 shows a case with $n = 2$, $p = 2$, in which II′ has no solution, and hence by Farkas' theorem I′ must have a solution.

2. The optimality criteria of linear programming: An application of Farkas' theorem

As a typical example of the use and power of the theorems of the alternative we shall show how Farkas' theorem can be employed to derive necessary optimality conditions for the following linear programming problem: Find an $\bar{x}$, if it exists, such that

1 $\quad -b\bar{x} = \min_{x \in X} (-bx) \qquad \bar{x} \in X = \{x \mid x \in R^n, \, Ax \leqq c\}$

where b and c are given fixed vectors in R^n and R^m respectively, and A is a given fixed $m \times n$ matrix.

2 $\qquad$ **Optimality criteria of linear programming**

(*Necessity*) *Let $\bar{x}$ be a solution of the linear programming problem 1. Then there exists a $\bar{u} \in R^m$ such that $(\bar{x}, \bar{u})$ satisfy*

3 $\quad A\bar{x} \leqq c \qquad$ (*dual feasibility*)†

4 $\quad A'\bar{u} = b$ ⎫

5 $\quad \bar{u} \geqq 0$ ⎭ $\quad$ (*primal feasibility*)†

6 $\quad b\bar{x} = c\bar{u} \qquad$ (*complementarity*)†

(*Sufficiency*) *If some $\bar{x} \in R^n$ and some $\bar{u} \in R^m$ satisfy 3 to 6, then $\bar{x}$ solves 1.*

PROOF

(*Necessity*) Let $\bar{x}$ solve 1. Define the index sets P, Q, and M as follows:

$P \cup Q = M = \{1, \ldots, m\} \qquad P = \{i \mid A_i\bar{x} = c_i\}$

$$Q = \{i \mid A_i\bar{x} < c_i\}$$

and assume that P and Q contain p and q elements, respectively. Then (see 1.4.16)

---

† These are standard terms of linear programming [Dantzig 63], which differ from the terminology of nonlinear programming (see Chap. 8). The complementarity condition usually refers to an equivalent form of 6: $\bar{u}(A\bar{x} - c) = 0$.

7    $A_P \bar{x} = c_P$

8    $A_Q \bar{x} < c_Q$

If $P = \emptyset$, it follows that

$$A\bar{x} - c < -\delta e$$

for some real number $\delta > 0$, where e is an m-vector of ones. Then for each $x \in R^n$, we can find a real number $\alpha > 0$ such that

$$A(\bar{x} + \alpha x) - c < -\delta e + \alpha A x \leqq 0$$

and hence $\bar{x} + \alpha x \in X$. Since $\bar{x}$ is the minimum solution of 1, we have that for each $x \in R^n$ there exists an $\alpha > 0$ such that

$$-b\bar{x} \leqq -b(\bar{x} + \alpha x)$$

Hence

$$bx \leqq 0 \qquad \text{for each } x \in R^n$$

which implies that $b = 0$.† By taking $\bar{u} = 0 \in R^m$, the relations 3 to 6 are satisfied because $b = 0$.

If $P \neq \emptyset$, then we assert that the system

9    $A_P x \leqq 0 \qquad bx > 0$

has no solution $x \in R^n$. For if 9 did have a solution x, say, then αx would also be a solution of 9 for each $\alpha > 0$. Now consider the point $\bar{x} + \alpha x$, where x is a solution of 9 and $\alpha > 0$. Then

10    $-b(\bar{x} + \alpha x) \qquad < -b\bar{x} \qquad \text{for } \alpha > 0$ (by 9)

11    $A_P(\bar{x} + \alpha x) - c_P \leqq 0 \qquad \text{for } \alpha > 0$ (by 9, 7)

12    $A_Q(\bar{x} + \alpha x) - c_Q \leqq -\delta e + \alpha A_Q x \leqq 0 \qquad \text{for some } \alpha > 0 \qquad$ (by 8)

where in 12 the first inequality follows by defining e as a q-vector of ones and

$$-\delta = \max_{i \in Q} (A_i \bar{x} - c_i) < 0$$

and the second inequality of 12 holds for some $\alpha > 0$ because $-\delta < 0$. But relations 10 to 12 imply that $\bar{x} + \alpha x \in X$ and $-b(\bar{x} + \alpha x) < -b\bar{x}$, which contradicts the assumption that $\bar{x}$ is a solution of 1. Hence 9 has no solution $x \in R^n$, and by Farkas' theorem 2.1.1 the system

13    $A'_P y = b \qquad y \geqq 0$

† To see this, take for a fixed $i \in M$, $x_i = b_i$, $x_{j \neq i} = 0$, then $bx \leqq 0$ implies that $(b_i)^2 \leqq 0$. Hence $b_i = 0$. Repeating this process for each $i \in M$, we get $b = 0$.

must have a solution $y \in R^p$.   If we let $0 \in R^q$, we have then

14 $A_P' y + A_Q' 0 = b \qquad y \geqq 0$

and

15 $c_P y + c_Q 0 = y c_P = y A_P \bar{x} = b \bar{x} \qquad$ (by *7*, *13*)

By defining $\bar{u} \in E^m$ such that $\bar{u} = (\bar{u}_P, \bar{u}_Q)$, where $\bar{u}_P = y \in R^p$, $\bar{u}_Q = 0 \in R^q$, condition *14* becomes conditions *4* and *5*, and condition *15* becomes condition *6*.   Condition *3* holds because $\bar{x} \in X$.   This completes the necessity proof.

(*Sufficiency*)   Let $\bar{x} \in R^n$ and $\bar{u} \in R^m$ satisfy *3* to *6*, and let $x \in X$.   By *3* we have that $\bar{x} \in X$.   Now

$$-bx - (-b\bar{x}) = -b(x - \bar{x})$$

$$= -\bar{u}A(x - \bar{x}) \qquad \text{(by *4*)}$$

$$= -\bar{u}Ax + c\bar{u} \qquad \text{(by *4*, *6*)}$$

$$= -\bar{u}(Ax - c)$$

$$\geqq 0 \qquad\qquad \text{(by *5*, $x \in X$)} \quad \blacksquare$$

We remark that Farkas' theorem was used only in establishing the necessity of the conditions *3* to *6*, whereas the sufficiency of the conditions *3* to *6* required only elementary arguments.   This is the typical situation in establishing optimality criteria in mathematical programming in general.   Necessity requires some fairly sophisticated mathematical tool, such as a theorem of the alternative or a separation theorem for convex sets (see Chap. 3), whereas sufficiency can be established merely by using elementary manipulation of inequalities.

The remaining sections of this chapter will be devoted to obtaining a rather comprehensive set of theorems of the alternative of the Farkas type.   We shall follow Tucker [Tucker 56] in establishing these results. We begin first in the next section with some existence theorems for linear inequalities, from which all the theorems of the alternative follow.

16   **Problem**

Consider the following general form of the linear programming problem: Find $\bar{x} \in R^n$, $\bar{y} \in R^\ell$, if they exist, such that

$$-b\bar{x} - a\bar{y} = \min_{x,y \in X} (-bx - ay)$$

$$(\bar{x},\bar{y}) \in X = \left\{ (x,y) \,\middle|\, \begin{array}{l} x \in R^n, y \in R^\ell \\ Ax + Dy \leq c \\ Bx + Ey = d \\ y \geq 0 \end{array} \right\}$$

where b, a, c, and d are given vectors in R^n, R^ℓ, R^m, and R^k respectively, and A, D, B, and E are given $m \times n$, $m \times \ell$, $k \times n$, and $k \times \ell$ matrices respectively. Show that a necessary and sufficient condition for $(\bar{x},\bar{y})$ to solve the above problem is that $(\bar{x},\bar{y})$ and some $\bar{u} \in R^m$ and $\bar{v} \in R^k$ satisfy the following conditions:

$$(\bar{x},\bar{y}) \quad \in X$$

$$A'\bar{u} + B'\bar{v} = b$$

$$D'\bar{u} + E'\bar{v} \geq a$$

$$\bar{u} \geq 0$$

$$c\bar{u} + d\bar{v} = b\bar{x} + a\bar{y}$$

(Hint: Replace the equality constraint

$$Bx + Ey = d$$

by

$$Bx + Ey \leq d$$

and

$$-Bx - Ey \leq -d$$

and use Theorem 2.)

3. Existence theorems for linear systems

We establish now some key theorems for the existence of certain types of solutions for linear systems and begin with a lemma due to Tucker [Tucker 56].

1   **Tucker's lemma**

For any given $p \times n$ matrix A, the systems

I $Ax \geqq 0$

and

II $A'y = 0, y \geqq 0$

possess solutions x and y satisfying

$A_1x + y_1 > 0$

PROOF   The proof is by induction on p. For $p = 1$, if $A_1 = 0$, take $y_1 = 1$, $x = 0$; if $A_1 \neq 0$, take $y_1 = 0$, $x = A_1$.

    Now assume that the lemma is true for a matrix A of p rows and proceed to prove it for a matrix of $p + 1$ rows $\bar{A}$:

$$\bar{A} = \begin{bmatrix} A \\ A_{p+1} \end{bmatrix} = \begin{bmatrix} A_1 \\ \cdot \\ \cdot \\ \cdot \\ A_p \\ A_{p+1} \end{bmatrix}$$

By applying the lemma to A, we get x,y satisfying

2 $Ax \geqq 0$ $A'y = 0$ $y \geqq 0$ $A_1x + y_1 > 0$

If $A_{p+1}x \geqq 0$, we take $\bar{y} = (y,0)$.   Then

3 $\bar{A}x \geqq 0$ $\bar{A}'\bar{y} = 0$ $\bar{y} \geqq 0$ $A_1x + y_1 > 0$

which extends the lemma to $\bar{A}$.

    However, if $A_{p+1}x < 0$, we apply the lemma a second time to the matrix B:

4 $$B = \begin{bmatrix} B_1 \\ \cdot \\ \cdot \\ \cdot \\ B_p \end{bmatrix} = \begin{bmatrix} A_1 + \lambda_1 A_{p+1} \\ \cdot \\ \cdot \\ \cdot \\ A_p + \lambda_p A_{p+1} \end{bmatrix}$$

where

5 $\lambda_j = \dfrac{A_j x}{-A_{p+1}x} \geqq 0$ $j = 1, \ldots , p$ (by *2*)

So

$$B_j x = A_j x + \lambda_j A_{p+1} x = 0$$

or

6 $Bx = 0$

This second use of the lemma yields v, u satisfying

7 $Bv \geq 0 \qquad B'u = 0 \qquad u \geq 0 \qquad B_1 v + u_1 > 0$

Let $\bar{u} = \left(u, \sum_{j=1}^{p} \lambda_j u_j \right)$.   It follows from 5 and 7 that

8 $\bar{u} \geq 0$

9 $\bar{A}' \bar{u} = A'u + A'_{p+1} \sum_{j=1}^{p} \lambda_j u_j = B'u - \sum_{j=1}^{p} \lambda_j A'_{p+1} u_j + A'_{p+1} \sum_{j=1}^{p} \lambda_j u_j = 0$

$$\text{(by } 4, 7)$$

Let

0 $w = v - \dfrac{A_{p+1} v}{A_{p+1} x} x$

then

1 $A_{p+1} w = A_{p+1} v - A_{p+1} v = 0$

and

2 $\bar{A} w = \begin{bmatrix} A \\ A_{p+1} \end{bmatrix} w = \begin{bmatrix} Aw \\ 0 \end{bmatrix} = \begin{bmatrix} A_1 w \\ \cdot \\ \cdot \\ \cdot \\ A_p w \\ 0 \end{bmatrix} = \begin{bmatrix} (B_1 - \lambda_1 A_{p+1}) w \\ \cdot \\ \cdot \\ \cdot \\ (B_p - \lambda_p A_{p+1}) w \\ 0 \end{bmatrix}$

$$= \begin{bmatrix} Bw \\ 0 \end{bmatrix} = \begin{bmatrix} Bv - \dfrac{A_{p+1} v}{A_{p+1} x} Bx \\ 0 \end{bmatrix} = \begin{bmatrix} Bv \\ 0 \end{bmatrix} \geq 0$$

where the last inequality follows from 7, the equality before from 6, the equality before from 10, the equality before from 11, and the equality before from 4.   Finally from 4, 11, 10, 6, and 7 we have

3 $A_1 w + u_1 = (B_1 - \lambda_1 A_{p+1}) w + u_1 = B_1 w + u_1$

$$= B_1 v - \dfrac{A_{p+1} v}{A_{p+1} x} B_1 x + u_1 = B_1 v + u_1 > 0$$

Relations 9, 8, 12, and 13 extend the lemma to $\bar{A}$.   ∎

From Tucker's lemma two important existence theorems follow. These theorems assert the existence of solutions of two linear systems that have a certain positivity property.

14   **First existence theorem [Tucker 56]**

For any given $p \times n$ matrix A, the systems

I $Ax \geqq 0$

and

II $A'y = 0,\ y \geqq 0$

possess solutions x and y satisfying

$Ax + y > 0$

PROOF   In Tucker's lemma the row A_1 played a special role. By renumbering the rows of A, any other row, say A_i, could have played the same role. Hence, by Tucker's lemma *1*, there exist $x^i \in R^n$, $y^i \in R^p$, $i = 1, \ldots , p$, such that

15
$$\left. \begin{array}{l} Ax^i \geqq 0 \\ A'y^i = 0,\ y^i \geqq 0 \\ A_i x^i + y_i^i > 0 \end{array} \right\} \quad i = 1, \ldots , p$$

Define

16 $\displaystyle x = \sum_{i=1}^p x^i \qquad y = \sum_{i=1}^p y^i$

Hence by *15*, we have that

$$Ax = \sum_{i=1}^p Ax^i \geqq 0$$

$$A'y = \sum_{i=1}^p A'y^i = 0$$

$$y = \sum_{i=1}^p y^i \geqq 0$$

and for $i = 1, \ldots , p$,

$$A_i x + y_i = \underbrace{A_i x^i + y_i^i}_{\substack{>0 \\ (\text{by } 15)}} + \underbrace{\sum_{\substack{k=1 \\ k \neq i}}^p (A_i x^k + y_i^k)}_{\substack{\geqq 0 \\ (\text{by } 15)}} > 0$$

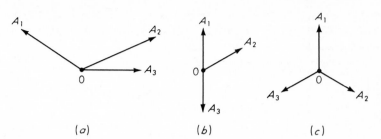

Fig. 2.3.1 Geometric interpretation of Theorem 14, $n = 2$, $p = 3$.

or

$$Ax + y > 0 \quad \blacksquare$$

A geometric interpretation of the above theorem can be given in the space of the rows $A_i \in R^n$, $i = 1, \ldots, p$. The theorem states that one of the following alternatives must occur for any given $p \times n$ matrix A:

(a) There exists a vector x which makes a strict acute angle ($<\pi/2$) with all the rows A_i, Fig. $2.3.1a$, or

(b) There exists a vector x which make an acute angle ($\leq \pi/2$) with all the rows A_i, and the origin 0 can be expressed as a nonnegative linear combination of the rows A_i with positive weights assigned to the rows A_i that are orthogonal to x, Fig. $2.3.1b$, or

(c) The origin can be expressed as a positive linear combination of the rows A_i, Fig. $2.3.1c$.

By letting the matrix A of the previous theorem have a special structure a second existence theorem can be easily established.

17   **Second existence theorem [Tucker 56]**

Let A and B be given $p^1 \times n$ and $p^2 \times n$ matrices, with A non-vacuous. Then the systems

I $Ax \geq 0$ $Bx = 0$

and

II $A'y_1 + B'y_2 = 0$, $y_1 \geq 0$

possess solutions $x \in R^n$, $y_1 \in R^{p^1}$, $y_2 \in R^{p^2}$ satisfying

$$Ax + y_1 > 0$$

PROOF   We remark first that the requirement that A be nonvacuous is merely to ensure that the statement of the theorem is nonvacuous, that is, without a matrix A the theorem says nothing.

We apply Theorem *14* to the systems

$$\begin{bmatrix} A \\ B \\ -B \end{bmatrix} x \geqq 0$$

and

$$[A', B', -B'] \begin{bmatrix} y_1 \\ z_1 \\ z_2 \end{bmatrix} = 0 \qquad \begin{bmatrix} y_1 \\ z_1 \\ z_2 \end{bmatrix} \geqq 0$$

and obtain x, y_1, z_1, z_2, satisfying

$$Ax + y_1 > 0$$

$$Bx + z_1 > 0$$

$$-Bx + z_2 > 0$$

Define now $y_2 = z_1 - z_2$. We have then that x, y_1, y_2 satisfy

$$Ax \geqq 0 \qquad Bx = 0$$

$$A'y_1 + B'y_2 = 0 \qquad y_1 \geqq 0$$

$$Ax + y_1 > 0 \quad \blacksquare$$

18   Corollary

Let A, B, C, and D be given $p^1 \times n$, $p^2 \times n$, $p^3 \times n$, and $p^4 \times n$ matrices, with A, B, or C nonvacuous. Then the systems

I $Ax \geqq 0 \qquad Bx \geqq 0 \qquad Cx \geqq 0 \qquad Dx = 0$

and

II $A'y_1 + B'y_2 + C'y_3 + D'y_4 = 0,\ y_1 \geqq 0,\ y_2 \geqq 0,\ y_3 \geqq 0$

possess solutions $x \in R^n$, $y_1 \in R^{p^1}$, $y_2 \in R^{p^2}$, $y_3 \in R^{p^3}$, $y_4 \in R^{p^4}$, satisfying

$$Ax + y_1 > 0$$

$$Bx + y_2 > 0$$

and

$$Cx + y_3 > 0$$

This corollary to Theorem *17* will be used to derive the most general types of the theorems of the alternative of the next section.

4.   Theorems of the alternative

In this section we shall be concerned with establishing a series of theorems relating to the certain occurrence of one of two mutually exclusive events. The two events, denoted by I and II, will be the existence of solutions of two related systems of linear inequalities and/or equalities.† The prototype of the theorem of the alternative can be stated as follows: Either I or II, but never both.   If we let $\bar{\text{I}}$ denote the nonoccurrence of I and similarly for $\overline{\text{II}}$, then we can state the following.

0   **Typical theorem of the alternative**

$\text{I} \Leftrightarrow \overline{\text{II}}$

or equivalently

$\bar{\text{I}} \Leftrightarrow \text{II}$

TYPICAL PROOF

$I \Rightarrow \overline{\text{II}}$ (or equivalently $\bar{\text{I}} \Leftarrow \text{II}$)

and

$\bar{\text{I}} \Rightarrow \text{II}$ (or equivalently $\text{I} \Leftarrow \overline{\text{II}}$)

The proof that $\text{I} \Rightarrow \overline{\text{II}}$ is usually quite elementary, but the proof that $\bar{\text{I}} \Rightarrow \text{II}$ utilizes the existence theorems of the previous section.

In the theorems to follow, certain obvious consistency conditions will not be stated explicitly for the sake of brevity.   For example, it will be understood that certain matrices must have the same number of rows, that the dimensionality of certain vectors must be the same as the number of columns in certain matrices, etc.

We begin now by establishing a fairly general theorem of the alternative due to Slater [Slater 51].

1   **Slater's theorem of the alternative [Slater 51]**

Let A, B, C, and D be given matrices, with A and B being nonvacuous. Then either

I $Ax > 0$ $Bx \geq 0$ $Cx \geqq 0$ $Dx = 0$ *has a solution x*

or

---

† Occasionally we shall also refer to the systems of inequalities and equalities themselves as systems I and II.

27

$$
\text{II} \quad \left\{ \begin{array}{l} A'y_1 + B'y_2 + C'y_3 + D'y_4 = 0 \\[4pt] \textit{with} \\[4pt] y_1 \geq 0,\ y_2 \geq 0,\ y_3 \geq 0 \qquad \textit{or} \\[4pt] y_1 \geq 0,\ y_2 > 0,\ y_3 \geq 0 \end{array} \right\} \textit{has a solution } y_1,\ y_2,\ y_3,\ y_4
$$

but never both.

PROOF (I $\Rightarrow \overline{\text{II}}$) By assumption, I holds. We will now show that if II also holds, then a contradiction ensues. If both I and II hold, then we would have x, y_1, y_2, y_3, y_4 such that

$$xA'y_1 + xB'y_2 + xC'y_3 + xD'y_4 > 0$$

because $xD'y_4 = 0$, $xC'y_3 \geq 0$, and either $xB'y_2 \geq 0$ and $xA'y_1 > 0$, or $xB'y_2 > 0$ and $xA'y_1 \geq 0$. But now we have a contradiction to the first equality of II. Hence I and II cannot hold simultaneously. Thus, I $\Rightarrow \overline{\text{II}}$.

($\overline{\text{I}} \Rightarrow$ II)

$$
\overline{\text{I}} \Rightarrow \left\{ \langle Ax \geq 0,\ Bx \geq 0,\ Cx \geq 0,\ Dx = 0 \rangle \Rightarrow \left\langle \begin{array}{l} Ax \not> 0 \\ \text{or} \\ Bx = 0 \end{array} \right\rangle \right\}
$$

$$
\Rightarrow \left\langle \left\langle \begin{array}{l} Ax \geq 0,\ Bx \geq 0,\ Cx \geq 0,\ Dx = 0 \\ A'y_1 + B'y_2 + C'y_3 + D'y_4 = 0 \\ y_1 \geq 0,\ y_2 \geq 0,\ y_3 \geq 0 \end{array} \right\rangle \Rightarrow \left\langle \begin{array}{l} y_1 \geq 0 \\ \text{or} \\ y_2 > 0 \end{array} \right\rangle \right\rangle
$$

$$\text{(by Corollary 18)}$$

$\Rightarrow$ II. ∎

We remark that in the above proof, the requirement that both A and B be nonvacuous was used essentially in establishing the fact that I $\Rightarrow \overline{\text{II}}$. Corollary *18*, which was used to prove that $\overline{\text{I}} \Rightarrow$ II, can handle systems in which merely A or B are nonvacuous. By slightly modifying the above proof, the cases B vacuous and A vacuous lead respectively to Motzkin's theorem of the alternative (or transposition theorem, as Motzkin called it) [Motzkin 36] and Tucker's theorem of the alternative [Tucker 56].

2   **Motzkin's theorem of the alternative [Motzkin 36]**

Let A, C, and D be given matrices, with A being *nonvacuous*.   *Then either*

I $\quad Ax > 0 \qquad Cx \geq 0 \qquad Dx = 0 \qquad$ *has a solution x*

or

II $\quad \left\langle \begin{array}{l} A'y_1 + C'y_3 + D'y_4 = 0 \\ \\ y_1 \geq 0, \, y_3 \geq 0 \end{array} \right\rangle$ *has a solution y_1, y_3, y_4*

but never both.

PROOF $\quad$ (I $\Rightarrow \overline{\text{II}}$) $\quad$ If both I and II hold, then we would have x, y_1, y_3, y_4 such that

$$xA'y_1 + xC'y_3 + xD'y_4 > 0$$

because $xD'y_4 = 0$, $xC'y_3 \geq 0$, and $xA'y_1 > 0$. But now we have a contradiction to the first equality of II. Hence, I and II cannot hold simultaneously. Thus, I $\Rightarrow \overline{\text{II}}$.

($\overline{\text{I}} \Rightarrow$ II)

$\overline{\text{I}} \Rightarrow \quad \Big\langle Ax \geq 0, \, Cx \geq 0, \, Dx = 0 \Big\rangle \Rightarrow \langle Ax \ngtr 0 \rangle \Big\rangle$

$\Rightarrow \Bigg\langle\!\!\Bigg\langle \left\langle \begin{array}{l} Ax \geq 0, \, Cx \geq 0, \, Dx = 0 \\ A'y_1 + C'y_3 + D'y_4 = 0 \\ y_1 \geq 0, \, y_3 \geq 0 \end{array} \right\rangle \Rightarrow \langle y_1 \geq 0 \rangle \Bigg\rangle\!\!\Bigg\rangle \qquad$ (by Corollary *18*)

$\Rightarrow$ II. $\quad\blacksquare$

3 $\qquad$ **Tucker's theorem of the alternative [Tucker 56]**

$\qquad$ *Let B, C, and D be given matrices, with B being nonvacuous. Then either*

I $\quad Bx \geq 0 \qquad Cx \geq 0 \qquad Dx = 0$ *has a solution x*

or

II $\quad \left\langle \begin{array}{l} B'y_2 + C'y_3 + D'y_4 = 0 \\ y_2 > 0, \, y_3 \geq 0 \end{array} \right\rangle$ *has a solution y_2, y_3, y_4*

but never both.

The proof is similar to the proof of Motzkin's theorem *2*. (The reader is urged to go through the proof himself here.)

$\qquad$ Slater [Slater 51] considered his theorem as the one providing the most general system I possible, because it involved all the ordering relations $>, \geq, \geqq, =$. Similarly we can derive another theorem which involves the most general system II possible, in which $y_1 > 0$, $y_2 \geq 0$, $y_3 \geqq 0$, and y_4 is unrestricted.

4   Theorem of the alternative

Let A, B, C, and D be given matrices, with A and B being nonvacuous. Then either

I $\quad \left\{ \begin{array}{l} Ax \geq 0,\ Bx \geqq 0,\ Cx \geqq 0,\ Dx = 0 \\ or \\ Ax \geqq 0,\ Bx > 0,\ Cx \geqq 0,\ Dx = 0 \end{array} \right\} \quad$ *has a solution x*

or

II $\quad \left\langle \begin{array}{l} A'y_1 + B'y_2 + C'y_3 + D'y_4 = 0 \\ y_1 > 0,\ y_2 \geqq 0,\ y_3 \geqq 0 \end{array} \right\rangle \quad$ *has a solution* $y_1,\ y_2,\ y_3,\ y_4$

but never both.

PROOF $\quad (\mathrm{I} \Rightarrow \overline{\mathrm{II}}) \quad$ If both I and II hold, then we would have x, y_1, y_2, y_3, y_4 satisfying

$$xA'y_1 + xB'y_2 + xC'y_3 + xD'y_4 > 0$$

because $xD'y_4 = 0$, $xC'y_3 \geqq 0$, and either $xB'y_2 \geqq 0$ and $xA'y_1 > 0$, or $xB'y_2 > 0$ and $xA'y_1 \geqq 0$. But now we have a contradiction to the first equality of II. Hence, I and II cannot hold simultaneously. Thus, $\mathrm{I} \Rightarrow \overline{\mathrm{II}}$.

$(\overline{\mathrm{II}} \Rightarrow \mathrm{I})$

$\overline{\mathrm{II}} \Rightarrow \left\langle \left\langle \begin{array}{l} A'y_1 + B'y_2 + C'y_3 + D'y_4 = 0 \\ \\ y_1 \geqq 0,\ y_2 \geqq 0,\ y_3 \geqq 0 \end{array} \right\rangle \Rightarrow \left\langle \begin{array}{l} y_1 \not> 0 \\ or \\ y_2 = 0 \end{array} \right\rangle \right\rangle$

$\Rightarrow \left\langle \left\langle \begin{array}{l} A'y_1 + B'y_2 + C'y_3 + D'y_4 = 0 \\ y_1 \geqq 0,\ y_2 \geqq 0,\ y_3 \geqq 0 \\ Ax \geqq 0,\ Bx \geqq 0,\ Cx \geqq 0,\ Dx = 0 \end{array} \right\rangle \Rightarrow \left\langle \begin{array}{l} Ax \geq 0 \\ or \\ Bx > 0 \end{array} \right\rangle \right\rangle$

(by Corollary *18*)

$\Rightarrow \mathrm{I}.$ ∎

We remark that if either A or B is vacuous, then we revert to Tucker's theorem *3* or Motzkin's theorem *2*.

We remark further that in all of the above theorems of the alternative the systems I are all homogeneous. Hence, by defining $z = -x$, the system I of, say, Slater's theorem *1* can be replaced by

I′ $\quad Az < 0,\ Bz \leqq 0,\ Cz \leqq 0,\ Dz = 0$ has a solution z

Similar remarks apply to theorems *2* through *4*.

The above theorems of the alternative subsume in essence all other theorems of this type. We derive below some of these theorems directly from the above ones.

5   ### Gordan's theorem [Gordan 73]

For each given matrix A, either

I $Ax > 0$ *has a solution* x

or

II $A'y = 0,\ y \geq 0$ *has a solution* y

but never both.

PROOF Follows directly from Motzkin's theorem *2*, by deleting the matrices C and D. ∎

Geometrically we may interpret Gordan's theorem as follows. Either there exists a vector x which makes a strict acute angle ($<\pi/2$) with all the row vectors of A, Fig. *2.4.1a*, or the origin can be expressed as a nontrivial, nonnegative linear combination of the rows of A, Fig. *2.4.1b*.

6   ### Farkas' theorem [Farkas 02]

For each given $p \times n$ matrix A and each given vector b in R^n either

I $Ax \leqq 0$ $bx > 0$ *has a solution* $x \in R^n$

or

II $A'y = b,\ y \geqq 0$ *has a solution* $y \in R^p$

but never both.

PROOF By Motzkin's theorem *2*, either I holds or

II′ $b\eta - A'y_3 = 0,\ \eta \geq 0,\ y_3 \geqq 0$ must have a solution $\eta \in R$ and

$$y_3 \in R^p$$

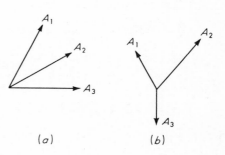

Fig. 2.4.1 Geometric interpretation of
Gordan's theorem. (a) (b)

but not both.   Since $\eta \in R$, $\eta \geq 0$ means $\eta > 0$.   Dividing through by η and letting $y = y_3/\eta$, we have that II′ is equivalent to *II*.   ∎

7   **Stiemke's theorem [Stiemke 15]**

For each given matrix B, either

I   *$Bx \geq 0$ has a solution x*

or

II   *$B'y = 0$, $y > 0$ has a solution y*

but never both.

PROOF Follows directly from Tucker's theorem *3*, by deleting C and D.   ∎

8   **Nonhomogeneous Farkas' theorem [Duffin 56]**

For a given $p \times n$ matrix A, given vectors $b \in R^n$, $c \in R^p$, and a given scalar β, either

I   *$bx > \beta$, $Ax \leq c$ has a solution $x \in R^n$*

or

II $\left. \begin{array}{l} A'y = b,\ cy \leq \beta,\ y \geq 0 \\ or \\ A'y = 0,\ cy < 0,\ y \geq 0 \end{array} \right\}$ *has a solution $y \in R^p$*

but never both.

PROOF   The system I is equivalent to

I′ $\left\langle \begin{array}{l} bx - \beta\xi > 0 \\ \qquad\quad \xi > 0 \\ -Ax + c\xi \geq 0 \end{array} \right\rangle$ has a solution $x \in R^n$, $\xi \in R$

By Motzkin's theorem *2*, then, either I′ holds or

II′ $\left\langle \begin{array}{l} \begin{bmatrix} b & 0 \\ -\beta & 1 \end{bmatrix} \begin{bmatrix} \eta_1 \\ \zeta_1 \end{bmatrix} + \begin{bmatrix} -A' \\ c \end{bmatrix} y_3 = 0 \\[1em] \begin{bmatrix} \eta_1 \\ \zeta_1 \end{bmatrix} \geq 0,\ y_3 \geq 0 \end{array} \right\rangle$ has a solution $\eta_1 \in R$, $\zeta_1 \in R$, $y_3 \in R^p$

but not both.

Now we have either $\eta_1 > 0$ or $\eta_1 = 0$ (and hence $\zeta_1 > 0$). By defining $y = y_3/\eta_1$ in the first case and $y = y_3$ in the second, we have that II$'$ is equivalent to

II$''$ $\left\{\begin{array}{l} b - A'y = 0,\ -\beta + cy = -\zeta_1 \leqq 0,\ y \geqq 0 \\ \text{or} \\ -A'y = 0,\ cy = -\zeta_1 < 0,\ y \geqq 0 \end{array}\right.$ has a solution $y \in R^p$

II$''$ is equivalent to II. ∎

9 **Gale's theorem for linear equalities [Gale 60]**

For a given $p \times n$ matrix A and given vector $c \in E^p$, either

I $\quad Ax = c$ *has a solution* $x \in R^n$

or

II $\quad A'y = 0,\ cy = 1$ *has a solution* $y \in R^p$

but never both.

PROOF The system I is equivalent to

I$'$ $\quad \xi > 0,\ -\xi c + Ax = 0$ has a solution $\xi \in R,\ x \in R^n$

By Motzkin's theorem *2*, either I$'$ holds or

II$'$ $\left\{\begin{array}{l} y_1 - cy_4 = 0 \\ A'y_4 = 0 \\ y_1 \geq 0 \end{array}\right.$ has a solution $y_1 \in R,\ y_4 \in R^p$

but not both. Since $y_1 \in R,\ y_1 > 0$. By defining $y = y_4/y_1$, II$'$ is equivalent to II. ∎

10 **Gale's theorem for linear inequalities ($\leqq$) [Gale 60]**

For a given $p \times n$ matrix A and given vector $c \in R^p$, either

I $\quad Ax \leqq c$ *has a solution* $x \in R^n$

or

II $\quad A'y = 0,\ cy = -1,\ y \geqq 0$ *has a solution* $y \in R^p$

but never both.

PROOF The system I is equivalent to

I$'$ $\quad \xi > 0,\ c\xi - Ax \geqq 0$ has a solution $\xi \in R,\ x \in R^n$

Table 2.4.1   Theorems of the Alternative†

1	$Ax > 0,\ Bx \geq 0,\ Cx \geqq 0,\ Dx = 0$ (A and B nonvacuous) (Slater)	$A'y_1 + B'y_2 + C'y_3 + D'y_4 = 0$ $y_1 \geq 0,\ y_2 \geqq 0,\ y_3 \geqq 0$ or $y_1 \geqq 0,\ y_2 > 0,\ y_3 \geqq 0$
2	$Ax > 0,\ Cx \geqq 0,\ Dx = 0$ (A nonvacuous) (Motzkin)	$A'y_1 + C'y_3 + D'y_4 = 0$ $y_1 \geq 0,\ y_3 \geqq 0$
3	$Bx \geq 0,\ Cx \geqq 0,\ Dx = 0$ (B nonvacuous) (Tucker)	$B'y_2 + C'y_3 + D'y_4 = 0$ $y_2 > 0,\ y_3 \geqq 0$
4	$Ax \geqq 0,\ Bx \geqq 0,\ Cx \geqq 0,\ Dx = 0$ or $Ax \geqq 0,\ Bx > 0,\ Cx \geqq 0,\ Dx = 0$ (A and B nonvacuous)	$A'y_1 + B'y_2 + C'y_3 + D'y_4 = 0$ $y_1 > 0,\ y_2 \geqq 0,\ y_3 \geqq 0$
5	$Ax > 0$ (Gordan)	$A'y = 0,\ y \geq 0$
6	$bx > 0,\ Ax \leqq 0$ (Farkas)	$A'y = b,\ y \geqq 0$
7	$Bx \geq 0$ (Stiemke)	$B'y = 0,\ y > 0$
8	$bx > \beta,\ Ax \leqq c$ (Nonhomogeneous Farkas)	$A'y = b,\ cy \leqq \beta,\ y \geqq 0$ or $A'y = 0,\ cy < 0,\ y \geqq 0$
9	$Ax = c$ (Gale)	$A'y = 0,\ cy = 1$
10	$Ax \leqq c$ (Gale)	$A'y = 0,\ cy = -1,\ y \geqq 0$
11	$Ax \leqq c$	$A'y = 0,\ cy = -1,\ y \geqq 0$ or $A'y = 0,\ cy \leqq 0,\ y > 0$

† No "or" appearing in the above table and in Problems *2.4.12* to *2.4.17* is an exclusive "or."

By Motzkin's theorem *2*, either I′ holds or

$$
\text{II}' \quad \left\langle \begin{array}{c} y_1 + cy_3 = 0 \\ -A'y_3 = 0 \\ y_1 \geq 0,\ y_3 \geqq 0 \end{array} \right\rangle \text{ has a solution } y_1 \in R,\ y_3 \in R^p
$$

but not both. By defining $y = y_3/y_1$, II follows from II′. ∎

11   Theorem for linear inequalities ($\leq$)

For a given $p \times n$ matrix A and a given vector $c \in R^p$, either

I $Ax \leq c$ *has a solution* $x \in R^n$

or

II $\left\langle \begin{array}{c} A'y = 0,\ cy = -1,\ y \geq 0 \\ \text{or} \\ A'y = 0,\ cy \leq 0,\ y > 0 \end{array} \right\rangle$ *has a solution* $y \in R^p$

but never both.

PROOF   The system I is equivalent to

I' $\xi > 0,\ c\xi - Ax \geq 0$ has a solution $\xi \in R,\ x \in R^n$

By Slater's theorem *1*, either I' holds or

II' $\left\langle \begin{array}{c} y_1 + cy_2 = 0 \\ A'y_2 = 0 \\ y_1 \geq 0,\ y_2 \geq 0 \\ \text{or} \\ y_1 \geq 0,\ y_2 > 0 \end{array} \right\rangle$ has a solution $y_1 \in R,\ y_2 \in R^p$

but not both.   If for the case when $y_1 \geq 0,\ y_2 \geq 0$, we set $y = y_2/y_1$, and for the case when $y_1 \geq 0,\ y_2 > 0$, we set $y = y_2$, then II is equivalent to II'. ∎

In the table above, Table *2.4.1*, we give a convenient summary of all the above theorems of the alternative.

Problems

By using any of the above theorems *1* to *11*, establish the validity of the following theorems of the alternative (*12* to *17*): Either I holds, or II holds, but never both, where I and II are given below.

12   I $Ax \leq c,\ x \geq 0$ has a solution x

II $A'y \geq 0,\ cy < 0,\ y \geq 0$ has a solution y [Gale 60]

13   I $Ax \leq 0,\ x \geq 0$ has a solution x

II $A'y \geq 0,\ y > 0$ has a solution y [Gale 60]

14   I $Ax < 0,\ x \geq 0$ has a solution x

II $A'y \geq 0,\ y \geq 0$ has a solution y [Gale 60]

15   I $Ax < 0,\ x > 0$ has a solution x

     II $A'y \geqq 0,\ y \geq 0$ has a solution y

16   I $\left\langle \begin{array}{l} Ax \leq 0,\ x \geqq 0 \text{ or} \\ Ax \leqq 0,\ x > 0 \end{array} \right\rangle$ has a solution x

     II $A'y \geq 0,\ y > 0$ has a solution y

17   I $Ax \leqq 0,\ x \geq 0$ has a solution x

     II $A'y > 0,\ y \geq 0$ has a solution y

18   **Mnemonic hint**

        In all the theorems of the alternative *1* to *17* above, which involve homogeneous inequalities and/or homogeneous equalities, the following correspondence between the ordering relations, $>$, $\geq$, $\geqq$, $=$, occurs:

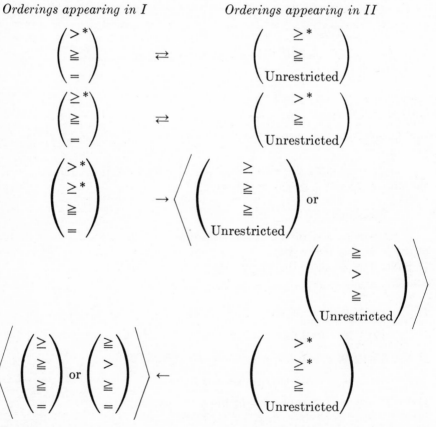

Orderings appearing in I   *Orderings appearing in II*

The asterisks indicate ordering relations which must be present in order for the correspondence to hold. The arrows indicate the direction in which the correspondence is valid; for example, $\rightarrow$ indicates that starting with the relations at the unpointed end of the arrow, the corresponding relations are those at the pointed end of the arrow.

Problem

Establish Motzkin's theorem 2 by starting with Farkas' theorem 6. (Hint: Let

$$\text{I} \Leftrightarrow \langle Ax > 0,\ Cx \geqq 0,\ Dx = 0 \text{ has a solution } x \in R^n \rangle$$

then show that

$$\bar{\text{I}} \Leftrightarrow \langle Ax \geqq e\varsigma,\ \varsigma > 0,\ Cx \geqq 0,\ Dx = 0 \text{ has no solution } x \in R^n,\ \varsigma \in R \rangle$$

and use Farkas' theorem. e is a vector of ones in the above.)

Chapter Three

Convex Sets in R^n

The purpose of this chapter is to introduce the fundamental concept of convex sets, to describe some properties of these sets, and to derive the basic separation theorems for convex sets. These separation theorems are the foundations on which many optimality conditions of nonlinear programming rest.

1. Convex sets and their properties

In order to define the concept of a convex set, we begin by defining line and line segments through two points in R^n.

1   Line

Let $x^1, x^2 \in R^n$. The *line* through x^1 and x^2 is defined as the set

$$\{x \mid x = (1 - \lambda)x^1 + \lambda x^2, \lambda \in R\}$$

or equivalently

$$\{x \mid x = p_1 x^1 + p_2 x^2, p_1, p_2 \in R, \\ p_1 + p_2 = 1\}$$

If we rewrite the first definition in the equivalent form

$$\{x \mid x = x^1 + \lambda(x^2 - x^1), \lambda \in R\}$$

and consider the case when $x \in R^2$, it becomes obvious that the vector equation $x = x^1 + \lambda(x^2 - x^1)$ is the parametric equation of elementary analytic geometry of the line through x^1 and x^2, Fig. *3.1.1*.

2   **Line segments**

Let $x^1, x^2 \in R^n$. We define the following *line segments* joining x^1 and x^2:

  (i)   Closed line segment $[x^1, x^2] = \{x \mid x = (1 - \lambda)x^1 + \lambda x^2, 0 \leq \lambda \leq 1\}$
  (ii)   Open line segment $(x^1, x^2) = \{x \mid x = (1 - \lambda)x^1 + \lambda x^2, 0 < \lambda < 1\}$
  (iii)   Closed-open line segment $[x^1, x^2) = \{x \mid x = (1 - \lambda)x^1 + \lambda x^2, 0 \leq \lambda < 1\}$
  (iv)   Open-closed line segment $(x^1, x^2] = \{x \mid x = (1 - \lambda)x^1 + \lambda x^2, 0 < \lambda \leq 1\}$

Obviously $[x^1, x^2]$ is the portion of the straight line through x^1 and x^2 which lies between and includes the points x^1 and x^2, Fig. *3.1.1*. (x^1, x^2) does not include x^1 or x^2, $[x^1, x^2)$ does not include x^2, and $(x^1, x^2]$ does not include x^1.

3   **Convex set**

A set $\Gamma \subset R^n$ is a *convex set* if the closed line segment† joining every two points of Γ is in Γ. Equivalently we have that a set $\Gamma \subset R^n$ is convex if

$$\left. \begin{array}{l} x^1, x^2 \in \Gamma \\[2mm] \lambda \in R, 0 \leq \lambda \leq 1 \end{array} \right\} \Rightarrow (1 - \lambda)x^1 + \lambda x^2 \in \Gamma$$

Figure *3.1.2* depicts some convex sets in R^2, and Fig. *3.1.3* some non-convex sets in R^2. It follows from *3* that R^n itself is convex, that the empty set is convex, and that all sets consisting each of one point are convex.

The subsets of R^n defined below in *4*, *5*, and *6* are all convex sets in R^n. This can be easily established by a direct verification of the definition *3* of a convex set.

† It is obvious that the definition of a convex set would be unchanged if any of the other line segments defined in *2* were used here instead of the closed line segment.

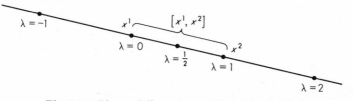

Fig. 3.1.1 Line and line segment through x^1 and x^2.

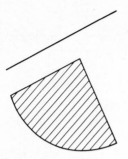

Fig. 3.1.2 Convex sets.

4   Halfspace

Let $c \in R^n$, $c \neq 0$, and $\alpha \in R$.   Then the set $\{x \mid x \in R^n, cx < \alpha\}$ is an *open halfspace* in R^n, and the set $\{x \mid x \in R^n, cx \leq \alpha\}$ is a *closed halfspace* in R^n.   (Both halfspaces are convex sets.)

5   Plane

Let $c \in R^n$, $c \neq 0$, and $\alpha \in R$.   Then the set $\{x \mid x \in R^n, cx = \alpha\}$ is called a *plane* in R^n.   (Each plane in R^n is a convex set.)

6   Subspace

A set $\Gamma \subset R^n$ is a *subspace* if

$$\left.\begin{array}{l} x^1, x^2 \in \Gamma \\ p_1, p_2 \in R \end{array}\right\} \Rightarrow p_1 x^1 + p_2 x^2 \in \Gamma$$

Each subspace of R^n contains the origin and is a convex set.   The subspaces of R^3 consist of $\emptyset$, R^3, the origin, and all straight lines and planes passing through the origin.

7   Problem

(i)   Show that each open or closed ball

$$B_\epsilon(\bar{x}) = \{x \mid x \in R^n, \|x - \bar{x}\| < \epsilon\} \quad \bar{B}_\epsilon(\bar{x}) = \{x \mid x \in R^n, \|x - \bar{x}\| \leq \epsilon\}$$

Fig. 3.1.3 Nonconvex sets.

around a point $\bar{x} \in R^n$ is a convex set.   (Hint: Use the triangle inequality *1.3.10* in the form $\|x + y\| \leqq \|x\| + \|y\|$.)

(ii)   Show that the interior of a convex set is convex.

8   Vertex

Let Γ be a convex set in R^n.   Each $x \in \Gamma$ for which there exist no two distinct $x^1, x^2 \in \Gamma$ different from x such that $x \in [x^1, x^2]$, is called a *vertex* of Γ (or an *extreme point* of Γ).

A convex set $\Gamma \subset R^n$ may have no vertices (for example the plane $\{x \mid x \in R^n, cx = \alpha\}$ and the open ball $B_\lambda(\bar{x})$ have no vertices), a finite number of vertices (for example the set $\{x \mid x \in R^n, x \geqq 0, ex = 1\}$, where e in an n-vector of ones, has the n vertices e^i, $i = 1, \ldots, n$, where e^i is an n-vector with $e_i{}^i = 1$ and $e_j{}^i = 0$, $i \neq j$), or an infinite number of vertices (for example the closed ball $\bar{B}_\lambda(\bar{x}) \subset R^n$ has an infinite number of vertices given by $\{x \mid x \in R^n, \|x - \bar{x}\| = \lambda\}$).

9   Theorem

If $(\Gamma_i)_{i \in I}$ is a family (finite or infinite) of convex sets in R^n, then their intersection $\bigcap\limits_{i \in I} \Gamma_i$ is a convex set.

PROOF   Let $x^1, x^2 \in \bigcap\limits_{i \in I} \Gamma_i$, and let $0 \leqq \lambda \leqq 1$.   Then for each $i \in I$, $x^1, x^2 \in \Gamma_i$, and since Γ_i is convex, $(1 - \lambda)x^1 + \lambda x^2 \in \Gamma_i$.   ∎

10   Polytope and polyhedron

A set in R^n which is the intersection of a finite number of *closed* halfspaces in R^n is called a *polytope*.   If a polytope is bounded (that is, for each x in the polytope $\|x\| \leqq \alpha$ for some fixed $\alpha \in R$), it is called a *polyhedron*.

It follows from the convexity of the halfspaces *4* and Theorem *9* that polytopes and polyhedra are convex sets.

11   Convex combination

A point $b \in R^n$ is said to be a *convex combination* of the vectors $a^1, \ldots, a^m \in R^n$ if there exist m real numbers $p_1, \ldots, p_m$ such that

$$b = p_1 a^1 + \cdots + p_m a^m, \; p_1, \ldots, p_m \geqq 0, \; p_1 + \cdots + p_m = 1$$

Equivalently, if we define an $m \times n$ matrix A whose ith row is $A_i = a^i$, and if we let $p = (p_1, \ldots, p_m) \in R^m$ and e be an m-vector of ones, then we have that b is a convex combination of the rows of A if

$$\langle b = A'p, \; p \geq 0, \; ep = 1 \rangle \text{ has a solution } p \in R^m$$

Note that if b is a convex combination of two points $a^1, a^2 \in R^n$, then this is equivalent to saying that $b \in [a^1, a^2]$ (see *2*).

12   **Simplex**

Let $x^0, x^1, \ldots, x^m$ be $m + 1$ distinct points in R^n, with $m \leq n$. If the vectors $x^1 - x^0, \ldots, x^m - x^0$ are linearly independent, then the set of all convex combinations of $x^0, x^1, \ldots, x^m$

$$S = \left\{ z \mid z = \sum_{i=0}^{m} p_i x^i, \; p_i \in R, \; p_i \geq 0, \; i = 0, \ldots, m, \; \sum_{i=0}^{m} p_i = 1 \right\}$$

is called an *m-simplex* in R^n with vertices $x^0, x^1, \ldots, x^m$. (A 0-simplex is a point, a 1-simplex is a closed line segment, a 2-simplex is a triangle, and a 3-simplex is a tetrahedron.)

13   **Theorem**

A set $\Gamma \subset R^n$ is convex if and only if for each integer $m \geq 1$, every convex combination of any m points of Γ is in Γ. Equivalently, a necessary and sufficient condition for the set Γ to be convex is that for each integer $m \geq 1$

14
$$\left. \begin{array}{l} x^1, \ldots, x^m \in \Gamma \\[4pt] p_1, \ldots, p_m \geq 0 \\[4pt] p_1 + \cdots + p_m = 1 \end{array} \right\} \Rightarrow p_1 x^1 + \cdots + p_m x^m \in \Gamma$$

PROOF   The sufficiency of *14* is trivial; take $m = 2$, then Γ is convex by *3*.

The necessity of *14* will be shown by induction. For $m = 1$, *14* holds trivially. For $m = 2$, *14* holds as a consequence of *3*. Assume now that *14* holds for m, we will now show that it also holds for $m + 1$. Let

$$x^1, x^2, \ldots, x^{m+1} \in \Gamma$$

$$p_1, \ldots, p_{m+1} \quad \geq 0$$

$$p_1 + \cdots + p_{m+1} = 1$$

If $p_{m+1} = 0$, then $p_1 x^1 + \cdots + p_m x^m \in \Gamma$, since *14* holds for m. If $p_{m+1} = 1$, then $p_1 x^1 + \cdots + p_{m+1} x^{m+1} = x^{m+1} \in \Gamma$. If $0 < p_{m+1} < 1$, then we can write

$$\sum_{i=1}^{m+1} p_i x^i = \left[\sum_{i=1}^{m} p_i\right]\left[\frac{p_1}{\sum\limits_{i=1}^{m} p_i} x^1 + \cdots + \frac{p_m}{\sum\limits_{i=1}^{m} p_i} x^m\right] + p_{m+1}x^{m+1} \in \Gamma \quad \blacksquare$$

(A point in Γ, because *14* holds for m)

(A point in Γ, because *14* holds for $m = 2$)

15 ## Carathéodory's theorem [Carathéodory 07]

Let $\Gamma \subset R^n$. If x is a convex combination of points of Γ, then x is a convex combination of $n + 1$ or fewer points of Γ.

PROOF Let

$$x = \sum_{i=1}^{m} p_i x^i, \ x^i \in \Gamma, \ p_i \in R, \ p_i \geqq 0, \ p_1 + \cdots + p_m = 1$$

We will show now that if $m > n + 1$, then x can be written as a convex combination of $m - 1$ points in Γ. (This would establish the theorem then, for we could repeatedly apply the result until x is a convex combination of $n + 1$ points of Γ.) If any p_i in the above expression is zero, then x is a convex combination of $m - 1$ or fewer points of Γ. So let each $p_i > 0$. Since $m > n + 1$, there exist $r_1, \ldots, r_{m-1} \in R$, not all zero, such that

$$r_1(x^1 - x^m) + \cdots + r_{m-1}(x^{m-1} - x^m) = 0 \qquad \text{(by } A.1.3)$$

Define $r_m = -(r_1 + \cdots + r_{m-1})$. Then

$$\sum_{i=1}^{m} r_i = 0 \qquad \sum_{i=1}^{m} r_i x^i = 0$$

Define

$$q_i = p_i - \alpha r_i \qquad \text{for } i = 1, \ldots, m$$

where α is some positive number chosen such that $q_i \geqq 0$ for all i, and at least one q_i, say q_k, is equal to 0. In particular we choose α such that

$$\frac{1}{\alpha} = \max_{i}\left\{\frac{r_i}{p_i}\right\} = \frac{r_k}{p_k}$$

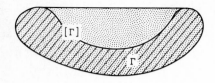

Fig. 3.1.4 A set Γ and its convex hull $[\Gamma]$.

Then

$$q_i \geqq 0, \, i = 1, \, \ldots , \, m, \, q_k = 0$$

$$\sum_{\substack{i=1 \\ i \neq k}}^{m} q_i = \sum_{i=1}^{m} q_i = \sum_{i=1}^{m} p_i - \alpha \sum_{i=1}^{m} r_i = \sum_{i=1}^{m} p_i = 1$$

and

$$x = \sum_{i=1}^{m} p_i x^i = \sum_{i=1}^{m} q_i x^i + \alpha \sum_{i=1}^{m} r_i x^i = \sum_{\substack{i=1 \\ i \neq k}}^{m} q_i x^i$$

Hence x is a convex combination of $m - 1$ points in Γ. ∎

16   **Convex hull**

Let $\Gamma \subset R^n$. The *convex hull* of Γ, denoted by $[\Gamma]$, is the intersection of all convex sets in R^n containing Γ. (By Theorem 9, the convex hull of any set $\Gamma \subset R^n$ is convex. Figure $3.1.4$ shows a hatched nonconvex set in R^2 and its shaded convex hull.)

Obviously if Γ is convex, then $\Gamma = [\Gamma]$.

17   **Theorem**

The convex hull $[\Gamma]$ of a set $\Gamma \subset R^n$ is equal to the set of all convex combinations of points of Γ.

PROOF   Let Λ denote the latter set, that is,

$$\Lambda = \left\{ x \mid x = \sum_{i=1}^{k} p_i a^i, \, p_i \in R, \, a^i \in \Gamma, \, p_i \geqq 0, \, \sum_{i=1}^{k} p_i = 1, \, k \geqq 1 \right\}$$

If $x^1, x^2 \in \Lambda$, then

$$x^1 = \sum_{i=1}^{k} p_i a^i, \, p_i \in R, \, a^i \in \Gamma, \, p_i \geqq 0, \, \sum_{i=1}^{k} p_i = 1$$

$$x^2 = \sum_{i=1}^{m} q_i b^i, \, q_i \in R, \, b^i \in \Gamma, \, q_i \geqq 0, \, \sum_{i=1}^{m} q_i = 1$$

Hence for $0 \leq \lambda \leq 1$

$$\lambda x^1 + (1 - \lambda)x^2 = \sum_{i=1}^{k} \lambda p_i a^i + \sum_{i=1}^{m} (1 - \lambda)q_i b^i$$

and

$$\lambda p_i \geq 0, \; (1 - \lambda)q_i \geq 0, \; \sum_{i=1}^{k} \lambda p_i + \sum_{i=1}^{m} (1 - \lambda)q_i = 1$$

Thus $\lambda x^1 + (1 - \lambda)x^2 \in \Lambda$, and Λ is convex. It is also clear that $\Gamma \subset \Lambda$. Since Λ is convex, then $[\Gamma] \subset \Lambda$. We also have by Theorem 13 that the convex set $[\Gamma]$ containing Γ must also contain all convex combinations of points of Γ. Hence $\Lambda \subset [\Gamma]$, and $\Lambda = [\Gamma]$. ∎

18   Sum of two sets

Let $\Gamma, \Lambda \subset R^n$. Their *sum* $\Gamma + \Lambda$ is defined by

$$\Gamma + \Lambda = \{z \mid z = x + y, \; x \in \Gamma, \; y \in \Lambda\}$$

19   Product of a set with a real number

Let $\Gamma \subset R^n$, and let $\lambda \in R$. The *product* $\lambda\Gamma$ is defined by

$$\lambda\Gamma = \{z \mid z = \lambda x, \; x \in \Gamma\}$$

Note that if $\lambda = -1$ and $\Gamma \subset \Lambda \subset R^n$, then $\Lambda + \lambda\Gamma = \Lambda - \Gamma$. Note that this is not the complement of Γ relative to Λ as defined in *1.2* and written as $\Lambda \sim \Gamma$.

20   Theorem

The sum $\Gamma + \Lambda$ of two convex sets Γ and Λ in R^n is a convex set.

PROOF   Let $z^1, z^2 \in \Gamma + \Lambda$, then $z^1 = x^1 + y^1$ and $z^2 = x^2 + y^2$, where $x^1, x^2 \in \Gamma$ and $y^1, y^2 \in \Lambda$. For $0 \leq \lambda \leq 1$

$$(1 - \lambda)z^1 + \lambda z^2 = (1 - \lambda)x^1 + \lambda x^2 + (1 - \lambda)y^1 + \lambda y^2 \in \Gamma + \Lambda$$

|◄──────────►| |◄──────────►|
(A point in   (A point in
Γ, by convex- Λ, by convexity of Γ)   ity of Λ)

Hence $\Gamma + \Lambda$ is convex. ∎

21   Theorem

The product $\mu\Gamma$ of a convex set Γ in R^n and the real number μ is a convex set.

$PROOF$ Let $z^1, z^2 \in \mu\Gamma$, then $z^1 = \mu x^1$, $z^2 = \mu x^2$, where $x^1, x^2 \in \Gamma$.   For $0 \leq \lambda \leq 1$

$$(1 - \lambda)z^1 + \lambda z^2 = \mu[(1 - \lambda)x^1 + \lambda x^2] \in \mu\Gamma \quad \blacksquare$$

$$|\blacktriangleleft \longrightarrow|$$

(A point in Γ,
by convexity
of Γ)

22   **Corollary**

If Γ and Λ are two convex sets in R^n, then $\Gamma - \Lambda$ is a convex set.

2.   Separation theorems for convex sets

It is intuitively plausible that if we had two disjoint convex sets in R^n, then we could construct a plane such that one set would lie on one side of the plane and the other set on the other side.   Despite its simplicity, this is a rather deep result and is not easy to prove.   One version of this result, the Hahn-Banach theorem, can be established by only using the vector space properties *1.3.3* of R^n and not the topological properties induced by the norm $\|x\|$ [Berge 63, Valentine 64].   We shall, however, use these topological properties of R^n (all summarized in Appendix B) in deriving the separation theorems for convex sets.   In particular our method of proof will make use of Gordan's theorem of the alternative *2.4.5* and the finite intersection theorem of compact sets *B.3.2* (iii). (Knowledge of the contents of Appendix B is assumed from here on.)

1   **Separating plane**

The plane $\{x \mid x \in R^n, cx = \alpha\}$, $c \neq 0$, is said to *separate (strictly separate)* two nonempty sets Γ and Λ in R^n if

$$x \in \Gamma \Rightarrow cx \leq \alpha \qquad (cx < \alpha)$$

$$x \in \Lambda \Rightarrow cx \geq \alpha \qquad (cx > \alpha)$$

If such a plane exists, the sets Γ and Λ are said to be *separable (strictly separable)*.

Figure *3.2.1* gives a simple illustration in R^2 of two sets in R^n which are separable, but which are neither disjoint nor convex.   It should be remarked that in general separability does not imply that the sets are disjoint (Fig. *3.2.1*), nor is it true in general that two disjoint sets are separable (Fig. *3.2.2*).   However, if the sets are nonempty, convex, and

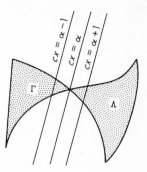

Fig. 3.2.1 Separable but not disjoint sets.

disjoint, then they are separable, and in fact this is a separation theorem we intend to prove.

Lemma

Let Ω be a nonempty convex set in R^n, not containing the origin 0. Then there exists a plane $\{x \mid x \in R^n, cx = 0\}$, $c \neq 0$, separating Ω and 0, that is,

$$x \in \Omega \Rightarrow cx \geq 0$$

PROOF With every $x \in \Omega$ we associate the nonempty closed set

$$\Lambda_x = \{y \mid y \in R^n, yy = 1, xy \geq 0\}$$

Let $x^1, \ldots, x^m$ be any finite set of points in Ω. It follows from the convexity of Ω, Theorem *3.1.13*, and from the fact that $0 \notin \Omega$, that

$$\sum_{i=1}^m x^i p_i = 0, \ \sum_{i=1}^m p_i = 1, \ p_i \geq 0, \ i = 1, \ldots, m \text{ has no solution } p \in R^m$$

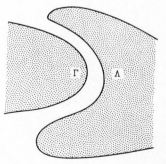

Fig. 3.2.2 Disjoint but not separable sets.

or equivalently

$$\sum_{i=1}^{m} x^i p_i = 0, \ p \geq 0 \text{ has no solution } p \in R^m$$

Hence by Gordan's theorem *2.4.5*

$x^i y > 0, \ i = 1, \ldots, m$ has a solution $y \in R^n$

Obviously $y \neq 0$, and we can take y such that $yy = 1$.   Then

$$y \in \bigcap_{i=1}^{m} \{y \mid y \in R^n, \ yy = 1, \ x^i y \geqq 0\} = \bigcap_{i=1}^{m} \Lambda_{x^i}$$

and hence

$$\bigcap_{i=1}^{m} \Lambda_{x^i} \neq \emptyset$$

The sets $(\Lambda_x)_{x \in \Omega}$ are closed sets relative to the compact set $\{y \mid y \in R^n,$ $yy = 1\}$ [see *B.1.8* and *B.3.2*(i)], hence by the finite intersection theorem *B.3.2*(iii) we have that $\bigcap_{x \in \Omega} \Lambda_x \neq \emptyset$.   Let c be any point in this intersection.   Then $cc = 1$ and $cx \geqq 0$ for all $x \in \Omega$.   Hence $\{x \mid x \in R^n,$ $cx = 0\}$ is the required separating plane. ∎

It should be remarked that in the above lemma we did not impose any conditions on Ω other than convexity.   The following example shows that the above lemma cannot be strengthened to $x \in \Omega \Rightarrow cx > 0$ without some extra assumptions.   The set

$$\Omega = \{x \mid x \in R^2, \ x \geq 0\} \cup \{x \mid x \in R^2, \ x_1 > 0, \ x_2 < 0\}$$

is convex and does not contain the origin, but there exists no plane $\{x \mid x \in R^n, \ cx = 0\}$ such that $x \in \Omega \Rightarrow cx > 0$ (Fig. *3.2.3*).

If on the other hand we do assume that Ω is closed (or even if we

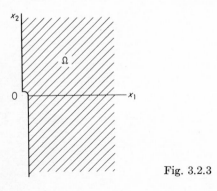

Fig. 3.2.3

assume less, namely that the origin is not a point of closure Ω), then we can establish a stronger result, that is, there exists a plane which strictly separates the origin from Ω (see Corollary 4 and Lemma 5 below). However, before doing this, we need to establish the following fundamental separation theorem.

3   **Separation theorem**

Let Γ and Λ be two nonempty disjoint convex sets in R^n. Then there exists a plane $\{x \mid x \in R^n, cx = \alpha\}, c \neq 0$, which separates them, that is,

$$x \in \Gamma \Rightarrow cx \leqq \alpha$$

$$x \in \Lambda \Rightarrow cx \geqq \alpha$$

PROOF   The set

$$\Lambda - \Gamma = \{x \mid x = y - z, y \in \Lambda, z \in \Gamma\}$$

is convex by Corollary *3.1.22*, and it does not contain the origin 0 because $\Gamma \cap \Lambda = \emptyset$. By Lemma *2* above there exists a plane $\{x \mid x \in R^n, cx = 0\}$, $c \neq 0$, such that

$$x \in \Lambda - \Gamma \Rightarrow cx \geqq 0$$

or

$$y \in \Lambda, z \in \Gamma \Rightarrow c(y - z) \geqq 0$$

Hence

$$\beta = \operatorname*{infimum}_{y \in \Lambda} cy \geqq \operatorname*{supremum}_{z \in \Gamma} cz = \gamma$$

Define

$$\alpha = \frac{\beta + \gamma}{2}$$

Then

$$z \in \Gamma \Rightarrow cz \leqq \alpha$$

$$y \in \Lambda \Rightarrow cy \geqq \alpha \quad \blacksquare$$

We derive now from the above fundamental separation theorem a corollary, and from the corollary a lemma, Lemma 5. Lemma 5 will be used in establishing a strict separation theorem, Theorem 6, below.

4   **Corollary**

Let Ω be a nonempty convex set in R^n. If the origin 0 is not a point of closure of Ω (or equivalently if the origin is not in the closure $\bar{\Omega}$ of Ω), then *there exists a plane $\{x \mid x \in R^n, cx = \alpha\}, c \neq 0, \alpha > 0$, strictly separating Ω and 0, and conversely. In other words*

$$\langle 0 \notin \bar{\Omega} \rangle \Leftrightarrow \left\langle \begin{array}{l} \exists c \neq 0, \alpha > 0: \\ x \in \Omega \Rightarrow cx > \alpha \end{array} \right\rangle$$

PROOF   ($\Leftarrow$)   Assume that there exist $c \neq 0, \alpha > 0$ such that $cx > \alpha$ for all $x \in \Omega$. If $0 \in \bar{\Omega}$, then (see *B.1.3* and *B.1.6*) there exists an $x \in \Omega$ such that $\|x\| < \alpha/2\|c\|$, and hence

$$\frac{\alpha}{2} = \|c\| \frac{\alpha}{2\|c\|} > \|c\| \cdot \|x\| \geq |cx| > \alpha \qquad \text{(by } 1.3.8\text{)}$$

which is a contradiction. Hence $0 \notin \bar{\Omega}$.

     ($\Rightarrow$)   Since 0 is not a point of closure of Ω, there exists an open ball $B_\epsilon(0) = \{x \mid x \in R^n, \|x\| < \epsilon\}$ around 0 such that $B_\epsilon(0) \cap \Omega = \emptyset$ (see *B.1.3*). Since the ball $B_\epsilon(0)$ is convex (see *3.1.7*), it follows by Theorem *3* that there exists a plane $\{x \mid x \in R^n, cx = \gamma\}, c \neq 0$, such that

$$x \in B_\epsilon(0) \Rightarrow cx \leq \gamma$$

$$x \in \Omega \Rightarrow cx \geq \gamma$$

Since $B_\epsilon(0)$ is an open ball, it must contain the nonzero vector δc for some positive δ. Hence $\gamma \geq \delta cc > 0$. Let $\alpha = \frac{1}{2}\delta cc > 0$. Then

$$x \in \Omega \Rightarrow cx \geq \gamma > \alpha > 0 \quad \blacksquare$$

5   **Lemma**

     Let Ω be a nonempty closed convex set in R^n. If Ω does not contain the origin, then there exists a plane $\{x \mid x \in R^n, cx = \alpha\}, c \neq 0, \alpha > 0$, strictly separating Ω and 0, and conversely. In other words

$$0 \notin \Omega \Leftrightarrow \left\langle \begin{array}{l} \exists c \neq 0, \alpha > 0: \\ x \in \Omega \Rightarrow cx > \alpha \end{array} \right\rangle$$

PROOF   This lemma follows from Corollary *4* above by observing that the requirement that Ω be closed and not contain the origin 0 implies that 0 is not a point of closure of Ω, that is, $0 \notin \bar{\Omega}$ (see *B.1.3, B.1.5* and *B.1.6*). $\blacksquare$

6   **Strict separation theorem**

     Let Γ and Λ be two nonempty convex sets in R^n, with Γ compact and Λ closed. If Γ and Λ are disjoint, then there exists a plane $\{x \mid x \in R^n,$

$cx = \alpha\}$, $c \neq 0$ *which strictly separates them, and conversely   In other words*

$$\Gamma \cap \Lambda = \emptyset \Leftrightarrow \left\langle \begin{array}{l} \exists c \neq 0 \text{ and } \alpha: \\ x \in \Gamma \Rightarrow cx < \alpha \\ x \in \Lambda \Rightarrow cx > \alpha \end{array} \right\rangle$$

PROOF ($\Leftarrow$)   If $x \in \Gamma \cap \Lambda$, then $cx < \alpha < cx$, a contradiction.
($\Rightarrow$)   The set

$$\Lambda - \Gamma = \{x \mid x = y - z, y \in \Lambda, z \in \Gamma\}$$

is convex by Corollary *3.1.22* and closed by Corollary *B.3.3*.   Hence by Lemma *5* above there exists a plane $\{x \mid x \in R^n, cx = \mu\}$, $c \neq 0$, $\mu > 0$, such that

$$x \in \Lambda - \Gamma \Rightarrow cx > \mu > 0$$

or

$$y \in \Lambda, z \in \Gamma \Rightarrow c(y - z) > \mu > 0$$

Hence

$$\beta = \inf_{y \in \Lambda} cy \geq \sup_{z \in \Gamma} cz + \mu > \sup_{z \in \Gamma} cz = \gamma$$

Define

$$\alpha = \frac{\beta + \gamma}{2}$$

Then

$$z \in \Gamma \Rightarrow cz < \alpha$$

$$y \in \Lambda \Rightarrow cy > \alpha \quad \blacksquare$$

The above separation theorems will be used to derive some fundamental theorems for convex functions in the next chapter, which in turn will be used in obtaining the fundamental Kuhn-Tucker saddlepoint optimality criteria of convex nonlinear programming in Chap. 5 and also the minimum principle necessary optimality condition of Chap. 11.

We remark here that a theorem of the alternative, the Gordan theorem *2.4.5*, was fundamental in deriving the above separation theorems.   We can reverse the process and use the above separation theorems to derive theorems of the alternative.   Thus to derive Gordan's theorem *2.4.5*, namely that either $A'y = 0$, $y \geq 0$ has a solution $y \in R^m$ or $Ax > 0$

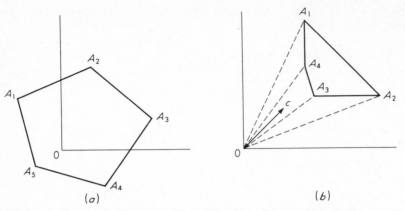

Fig. 3.2.4 Geometric reinterpretation of Gordan's theorem by using Lemma 5.
(a) $A'y = 0$, $y \geq 0$ has solution; $Ax > 0$ has no solution;
(b) $Ax > 0$ has solution; $A'y = 0$, $y \geq 0$ has no solution.

has a solution $x \in R^n$, we observe that if $e \in R^m$ is a vector of ones, then

$$A'y = 0, y \geq 0 \text{ has no solution} \Leftrightarrow 0 \notin \Omega = \{z \mid z = A'y, y \geq 0, ey = 1\}$$

$$\Leftrightarrow \exists c \neq 0, \alpha > 0: z \in \Omega \Rightarrow cz > \alpha$$

$$(\text{by } 5)$$

$$\Leftrightarrow y \geq 0, ey = 1 \Rightarrow cA'y > \alpha > 0$$

$$\Leftrightarrow Ac > 0$$

The last implication follows by taking $y = e^i \in R^m$, $i = 1, \ldots, m$, where e^i has zeros for all elements except 1 for the ith element.

Using the framework of Lemma 5 we can give a geometric reinterpretation of the Gordan's theorem as follows: Either the origin $0 \in R^n$

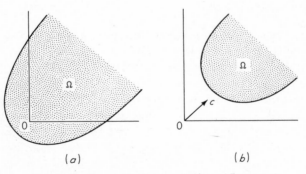

Fig. 3.2.5 Geometric interpretation of Lemma 5

is in the convex hull of the row vectors $A_1, \ldots, A_n$ of the matrix A ($A'y = 0$, $y \geq 0$ has a solution, Fig. *3.2.4a*), or it is not (in which case by Lemma *5* $Ax > 0$ has a solution $x = c$, Fig. *3.2.4b*). More generally, if Ω is any nonempty closed convex set in R^n, then either it contains the origin, Fig. *3.2.5a*, or it does not (in which case by Lemma *5* there exists a vector $c \in R^n$ which makes a strict acute angle with each $x \in \Omega$, Fig. *3.2.5b*).

7   ### Problem

Establish Farkas' theorem *2.4.6* by using Theorem *6* above. (Hint: Observe that $A'y = b$, $y \geq 0$ has no solution if and only if the sets $\{b\}$ and $\{z \mid z = A'y, y \geq 0\}$ are disjoint. Then use Theorem *6*.)

Chapter Four

Convex and Concave Functions

In this chapter we introduce convex, concave, strictly convex, and strictly concave functions defined on subsets of R^n. Convex and concave functions are extremely important in nonlinear programming because they are among the few functions for which sufficient optimality criteria can be given (Chaps. 5 and 7), and they are the only functions for which necessary optimality conditions can be given without linearization (Kuhn-Tucker saddlepoint condition in Chap. 5). We give in this chapter some of the basic properties of convex and concave functions and obtain some fundamental theorems involving these functions. These theorems, derived by using the separation theorems for convex sets of Chap. 3, are akin to the theorems of the alternative derived in Chap. 2 for linear systems. In this sense convex and concave functions inherit some of the important properties of linear functions. These fundamental theorems will be used to derive the important saddlepoint necessary optimality condition of Chap. 5 and the minimum principle necessary optimality condition of Chap. 11. Finally it should be mentioned that no differentiability or explicit continuity requirements are made on the functions introduced in this chapter. A subsequent chapter, Chap. 6, will be devoted to differentiable convex and concave functions.

1.   Definitions and basic properties

Convex function

1

A numerical function θ defined on a set $\Gamma \subset R^n$ is said to be *convex at* $\bar{x} \in \Gamma$ (with respect to Γ) if

$$\left.\begin{array}{l} x \in \Gamma \\ 0 \leqq \lambda \leqq 1 \\ (1 - \lambda)\bar{x} + \lambda x \in \Gamma \end{array}\right\} \Rightarrow (1 - \lambda)\theta(\bar{x}) + \lambda\theta(x) \geqq \theta[(1 - \lambda)\bar{x} + \lambda x]$$

θ is said to be *convex on* Γ if it is convex at each $x \in \Gamma$.

Note that this definition of a convex function is slightly more general than the customary definition in the literature [Fenchel 53, Valentine 64, Berge-Ghouila Houri 65] in that (i) we define convexity at a point first and then convexity on a set, and (ii) we do not require Γ to be a convex set.   This generalization will allow us to handle a somewhat wider class of problems later.   It follows immediately from the above definition that a numerical function θ defined on a convex set Γ is convex on Γ if and only if

$$\left.\begin{array}{l} x^1, x^2 \in \Gamma \\ 0 \leqq \lambda \leqq 1 \end{array}\right\} \Rightarrow (1 - \lambda)\theta(x^1) + \lambda\theta(x^2) \geqq \theta[(1 - \lambda)x^1 + \lambda x^2]$$

Figure *4.1.1* depicts two convex functions on convex subsets of $R^n = R$.

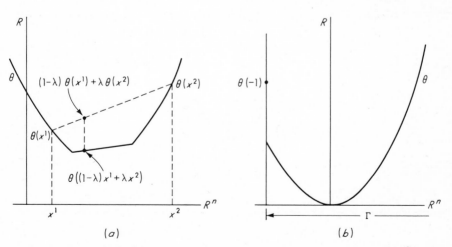

Fig. 4.1.1 Convex functions on subsets of $R^n = R$.   (*a*) A convex function θ on R; (*b*) A convex function θ on $\Gamma = [-1, \infty)$.

2   Concave function

A numerical function θ defined on a set $\Gamma \subset R^n$ is said to be *concave at* $\bar{x} \in \Gamma$ (with respect to Γ) if

$$\left.\begin{array}{c} x \in \Gamma \\ 0 \leq \lambda \leq 1 \\ (1 - \lambda)\bar{x} + \lambda x \in \Gamma \end{array}\right\} \Rightarrow (1 - \lambda)\theta(\bar{x}) + \lambda\theta(x) \leq \theta[(1 - \lambda)\bar{x} + \lambda x]$$

θ is said to be *concave on* Γ if it is concave at each $x \in \Gamma$.

Obviously θ is concave at $\bar{x} \in \Gamma$ (concave on Γ) if and only if $-\theta$ is convex at $\bar{x}$ (convex on Γ).   Results obtained for convex functions can be changed into results for concave functions by the appropriate multiplication by -1, and vice versa.

It follows immediately from the above definition that a numerical function θ defined on a convex set Γ is concave on Γ if and only if

$$\left.\begin{array}{c} x^1, x^2 \in \Gamma \\ 0 \leq \lambda \leq 1 \end{array}\right\} \Rightarrow (1 - \lambda)\theta(x^1) + \lambda\theta(x^2) \leq \theta[(1 - \lambda)x^1 + \lambda x^2]$$

Figure *4.1.2* depicts two concave functions on convex subsets of $R^n = R$.

3   Problem

Show that a linear function, $\theta(x) = cx - \alpha$, $x \in R^n$, is both convex and concave on R^n, and conversely.

4   Strictly convex function

A numerical function θ defined on a set $\Gamma \subset R^n$ is said to be *strictly*

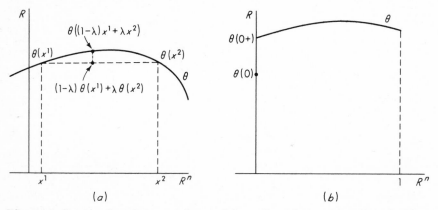

Fig. 4.1.2 Concave functions on subsets of $R^n = R$.   (a) A concave function θ on R; (b) A concave function θ on $\Gamma = [0,1]$.

convex at $\bar{x} \in \Gamma$ (with respect to Γ) if

$$\left.\begin{array}{c} x \in \Gamma \\ x \neq \bar{x} \\ 0 < \lambda < 1 \\ (1 - \lambda)\bar{x} + \lambda x \in \Gamma \end{array}\right\} \Rightarrow (1 - \lambda)\theta(\bar{x}) + \lambda\theta(x) > \theta[(1 - \lambda)\bar{x} + \lambda x]$$

θ is said to be *strictly convex on* Γ if it is strictly convex at each $x \in \Gamma$.

5   **Strictly concave function**

A numerical function θ defined on a set $\Gamma \subset R^n$ is said to be *strictly concave at* $\bar{x} \in \Gamma$ (with respect to Γ) if

$$\left.\begin{array}{c} x \in \Gamma \\ x \neq \bar{x} \\ 0 < \lambda < 1 \\ (1 - \lambda)\bar{x} + \lambda x \in \Gamma \end{array}\right\} \Rightarrow (1 - \lambda)\theta(\bar{x}) + \lambda\theta(x) < \theta[(1 - \lambda)\bar{x} + \lambda x]$$

θ is said to be *strictly concave on* Γ if it is strictly concave at each $x \in \Gamma$.

Obviously a strictly convex (strictly concave) function on a set $\Gamma \subset R^n$ is convex (concave) on Γ, but not conversely. For example a constant function on R^n is both convex and concave on R^n, but neither strictly convex nor strictly concave on R^n. In fact, it can be easily shown that all linear functions $\theta(x) = cx - \alpha$ on R^n are neither strictly convex nor strictly concave on R^n. Hence, because of the linear portion, the function depicted in Fig. *4.1.1a* is not strictly convex on R, but the function of Fig. *4.1.1b* is strictly convex on $[-1, \infty)$. Both functions of Fig. *4.1.2* are strictly concave on their domains of definition.

An n-dimensional vector function f defined on a set Γ in R^n is convex at $\bar{x} \in \Gamma$, convex on Γ, etc., if each of its components f_i, $i = 1$, . . . , m, is convex at $\bar{x} \in \Gamma$, convex on Γ, etc.

6   **Theorem**

Let $f = (f_1, \ . \ . \ . \ , f_m)$ *be an* m-*dimensional vector function defined on* $\Gamma \subset R^n$. *If* f *is convex at* $\bar{x} \in \Gamma$ (*convex on* Γ), *then each nonnegative linear combination of its components* f_i

$$\theta(x) = pf(x) \qquad p \geq 0$$

is convex at $\bar{x}$ (*convex on* Γ).

PROOF   Let $x \in \Gamma$, $0 \leqq \lambda \leqq 1$, and let $(1 - \lambda)\bar{x} + \lambda x \in \Gamma$.   Then

$$\theta[(1 - \lambda)\bar{x} + \lambda x] = pf[(1 - \lambda)\bar{x} + \lambda x]$$

$$\leqq p[(1 - \lambda)f(\bar{x}) + \lambda f(x)]$$

$$\text{(by convexity of } f \text{ at } \bar{x} \text{ and } p \geqq 0)$$

$$= (1 - \lambda)pf(\bar{x}) + \lambda pf(x)$$

$$= (1 - \lambda)\theta(\bar{x}) + \lambda\theta(x) \quad \blacksquare$$

7   **Problem**

Let θ be a numerical function defined on a convex set $\Gamma \subset R^n$. Show that θ is respectively convex, concave, strictly convex, or strictly concave on Γ if and only if for each $x^1, x^2 \in \Gamma$, the numerical function ψ defined on the line segment $[0,1]$ by

$$\psi(\lambda) = \theta[(1 - \lambda)x^1 + \lambda x^2]$$

is respectively convex, concave, strictly convex, or strictly concave on $[0,1]$.

8   **Theorem**

For a numerical function θ defined on a convex set $\Gamma \subset R^n$ to be convex on Γ it is necessary and sufficient that its epigraph

$$G_\theta = \{(x,\zeta) \mid x \in \Gamma, \zeta \in R, \theta(x) \leqq \zeta\} \subset R^{n+1}$$

be a convex set in R^{n+1}.

PROOF

(Sufficiency)   Assume that G_θ is convex.   Let $x^1, x^2 \in \Gamma$, then $[x^1, \theta(x^1)] \in G_\theta$ and $[x^2, \theta(x^2)] \in G_\theta$.   By the convexity of G_θ we have that

$$[(1 - \lambda)x^1 + \lambda x^2, (1 - \lambda)\theta(x^1) + \lambda\theta(x^2)] \in G_\theta \qquad \text{for } 0 \leqq \lambda \leqq 1$$

or

$$\theta[(1 - \lambda)x^1 + \lambda x^2] \leqq (1 - \lambda)\theta(x^1) + \lambda\theta(x^2) \qquad \text{for } 0 \leqq \lambda \leqq 1$$

and hence θ is convex on Γ.

(Necessity)   Assume that θ is convex on Γ.   Let $x^1, \zeta^1 \in G_\theta$ and $x^2, \zeta^2 \in G_\theta$. By the convexity of θ on Γ we have that for $0 \leqq \lambda \leqq 1$

$$\theta[(1 - \lambda)x^1 + \lambda x^2] \leqq (1 - \lambda)\theta(x^1) + \lambda\theta(x^2)$$

$$\leqq (1 - \lambda)\zeta^1 + \lambda\zeta^2$$

Hence

$$[(1 - \lambda)x^1 + \lambda x^2, (1 - \lambda)\zeta^1 + \lambda\zeta^2] \in G_\theta$$

and G_θ is a convex set in R^{n+1}. ∎

9 **Corollary**

For a numerical function θ defined on a convex set $\Gamma \subset R^n$ to be concave on Γ it is necessary and sufficient that its hypograph

$$H_\theta = \{(x,\zeta) \mid x \in \Gamma, \zeta \in R, \theta(x) \geqq \zeta\} \subset R^{n+1}$$

be convex set in R^{n+1}.

Figure *4.1.3a* depicts a convex function on Γ and its convex epigraph G_θ. Figure *4.1.3b* depicts a concave function on Γ and its convex hypograph H_θ.

10 **Theorem**

Let θ be a numerical function defined on a convex set $\Gamma \subset R^n$. A necessary but not sufficient condition for θ to be convex on Γ is that the set

$$\Lambda_\alpha = \{x \mid x \in \Gamma, \theta(x) \leqq \alpha\} \subset \Gamma \subset R^n$$

be convex for each real number α.

PROOF Let θ be convex on Γ and let $x^1, x^2 \in \Lambda_\alpha$. Then

$$\theta[(1 - \lambda)x^1 + \lambda x^2] \leqq (1 - \lambda)\theta(x^1) + \lambda\theta(x^2) \qquad \text{(by convexity of } \theta\text{)}$$

$$\leqq (1 - \lambda)\alpha + \lambda\alpha \qquad \text{(because } x^1, x^2 \in \Lambda_\alpha\text{)}$$

$$= \alpha$$

Hence $(1 - \lambda)x^1 + \lambda x^2 \in \Lambda_\alpha$, and Λ_α is convex.

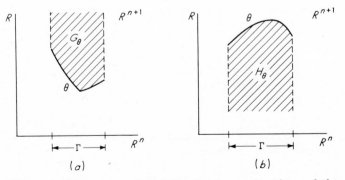

Fig. 4.1.3 The convex epigraph G_θ of a convex function and the convex hypograph H_θ of a concave function.

We now show that if Λ_α is convex for each α, it does not follow that θ is a convex function on Γ. Consider the function θ on R defined by $\theta(x) = (x)^3$. θ is not convex on R. However, the set

$$\Lambda_\alpha = \{x \mid x \in R, (x)^3 \leqq \alpha\} = \{x \mid x \in R, x \leqq (\alpha)^{\frac{1}{3}}\}$$

is obviously convex for any α (see 3.1.4). ∎

11   Corollary

Let θ be a numerical function defined on the convex set $\Gamma \subset R^n$. A necessary but not sufficient condition for θ to be concave on Γ is that the set

$$\Omega_\alpha = \{x \mid x \in \Gamma, \theta(x) \geqq \alpha\} \subset \Gamma \subset R^n$$

be convex for each real number α.

Figure 4.1.4a depicts a convex function θ on a convex set $\Gamma \subset R^n = R$ and the associated convex set Λ_α. Figure 4.1.4b depicts a nonconvex function θ and the associated convex set Λ_α. Figure 4.1.4c depicts a concave function θ on a convex set $\Gamma \subset R^n = R$ and the associated convex set Ω_α.

12   Problem

Let θ be a numerical function defined on the convex set $\Gamma \subset R^n$. Show that a necessary and sufficient condition for θ to be convex on Γ is that for each integer $m \geqq 1$

$$\left. \begin{array}{l} x^1, \ldots, x^m \in \Gamma \\ p_1, \ldots, p_m \geqq 0 \\ p_1 + \cdots + p_m = 1 \end{array} \right\} \Rightarrow \theta(p_1 x^1 + \cdots + p_m x^m) \leqq p_1 \theta(x^1) + \cdots + p_m \theta(x^m)$$

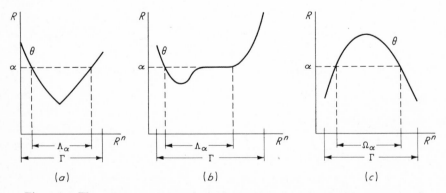

Fig. 4.1.4 The convex sets Λ_α and Ω_α of 10 and 11 associated with a function θ.

(Hint: Use Theorems *8* and *3.1.13*. The above inequality is Jensen's inequality [Jensen 06].)

13

Theorem

If $(\theta_i)_{i \in I}$ is a family (finite or infinite) of numerical functions which are convex and bounded from above on a convex set $\Gamma \subset R^n$, then the numerical function

$$\theta(x) = \sup_{i \in I} \theta_i(x)$$

is a convex function on Γ.

PROOF Since each θ_i is a convex function on Γ, their epigraphs

$$G_{\theta_i} = \{(x,\varsigma) \mid x \in \Gamma, \varsigma \in R, \theta_i(x) \leq \varsigma\}$$

are convex sets in R^{n+1} by Theorem *8*, and hence their intersection

$$\bigcap_{i \in I} G_{\theta_i} = \{(x,\varsigma) \mid x \in \Gamma, \varsigma \in R, \theta_i(x) \leq \varsigma, \forall i \in I\}$$

$$= \{(x,\varsigma) \mid x \in \Gamma, \varsigma \in R, \theta(x) \leq \varsigma\}$$

is also a convex set in R^{n+1} by Theorem *3.1.9*. But this convex intersection is the epigraph of θ. Hence θ is a convex function on Γ by Theorem *8*. ∎

14

Corollary

If $(\theta_i)_{i \in I}$ is a family (finite or infinite) of numerical functions which are concave and bounded from below on a convex set $\Gamma \subset R^n$, then the numerical function

$$\theta(x) = \inf_{i \in I} \theta_i(x)$$

is a concave function on Γ.

We end this section by remarking that a function θ which is convex on a convex set $\Gamma \subset R^n$ is not necessarily a continuous function. For example on the halfline $\Gamma = \{x \mid x \in R, x \geq -1\}$, the numerical function

$$\theta(x) = \begin{cases} 2 & \text{for } x = -1 \\ (x)^2 & \text{for } x > -1 \end{cases}$$

is a convex function on Γ, but is obviously not continuous at $x = -1$, Fig. *4.1.1b*. However, if Γ is an *open* convex set, then a convex function θ on Γ is indeed continuous. This fact is established in the following theorem.

15   Theorem

Let Γ be an open convex set in R^n. If θ is a convex numerical function on Γ then θ is continuous on Γ.

PROOF   [Fleming 65]†   Let $x^0 \in \Gamma$, and let α be the distance (see *1.3.9*) from x^0 to the closest point in R^n not in Γ ($\alpha = +\infty$ if $\Gamma = R^n$).   Let C be an n-cube with center x^0 and side length 2δ, that is

$$C = \{x \mid x \in R^n, \; -\delta \leqq x_i - x_i^0 \leqq \delta, \; i = 1, \ldots, n\}$$

By letting $(n)^{\frac{1}{2}}\delta < \alpha$, we have that $C \subset \Gamma$.   Let V denote the set of 2^n vertices of C.   Let

$$\beta = \max_{x \in V} \theta(x)$$

By Theorem *10* the set $\Lambda_\beta = \{x \mid x \in \Gamma, \theta(x) \leqq \beta\}$ is convex.   Since C is the convex hull of V (this can be easily shown by induction on n) and $V \subset \Lambda_\beta$, it follows that $C \subset \Lambda_\beta$, by Theorem *3.1.13* (Fig. *4.1.5*).

Let x be any point such that $0 < \|x - x^0\| < \delta$, and define $x^0 + u$, $x^0 - u$ on the line through x^0 and x as in Fig. *4.1.5*.   Write x now as a convex combination of x^0 and $x^0 + u$, and x^0 as a convex combination of x and $x^0 - u$.   If $\lambda = \|x - x^0\|/\delta$, then

$$x = x^0 + \lambda u = \lambda(x^0 + u) + (1 - \lambda)x^0$$

$$x^0 = x - \lambda u = x + \lambda(x^0 - u) - \lambda x^0$$

$$= \frac{1}{1 + \lambda} x + \frac{\lambda}{1 + \lambda} (x^0 - u)$$

† Fleming attributes this proof to F. J. Almgren.

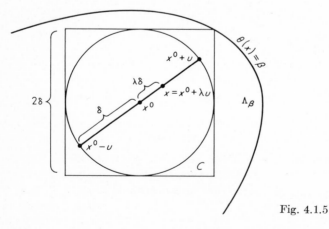

Fig. 4.1.5

Since θ is convex on Γ

$$\theta(x) \leq \lambda\theta(x^0 + u) + (1 - \lambda)\theta(x^0) \leq \lambda\beta + (1 - \lambda)\theta(x^0)$$

$$\theta(x^0) \leq \frac{1}{1 + \lambda}\,\theta(x) + \frac{\lambda}{1 + \lambda}\,\theta(x^0 - u) \leq \frac{\theta(x) + \lambda\beta}{1 + \lambda}$$

These inequalities give

$$-\lambda[\beta - \theta(x^0)] \leq \theta(x) - \theta(x^0) \leq \lambda[\beta - \theta(x^0)\}$$

or

$$|\theta(x) - \theta(x^0)| \leq \frac{\beta - \theta(x^0)}{\delta}\,\|x - x^0\|$$

Thus for any given $\epsilon > 0$ it follows that $|\theta(x) - \theta(x^0)| < \epsilon$ for all x satisfying $[\beta - \theta(x^0)]\,\|x - x^0\| < \delta$, and hence $\theta(x)$ is continuous at x^0. ∎

Since the interior of each set $\Gamma \subset R^n$ is open, it follows that if θ is a convex function on a convex set $\Gamma \subset R^n$, it is continuous on its interior.

2. Some fundamental theorems for convex functions

We saw in Chap. 2 that Farkas' theorem of the alternative played a crucial role in deriving the necessary optimality conditions of linear programming. In this section we shall derive what may be considered extensions of theorems of the alternative of Chap. 2 to convex and concave functions. These theorems in turn will play a similar crucial role in deriving the necessary optimality conditions of nonlinear programming in Chaps. 5 and 11. (In the remainder of this chapter various properties of continuous and semicontinuous functions will be used. For convenience, these results are summarized in Appendix C.)

We begin by establishing a fundamental theorem for convex functions, the essence of which is given in [Fan-Glicksburg-Hoffman 57].

1   **Theorem**

Let Γ be a nonempty convex set in R^n, let f be an m-dimensional convex vector function on Γ, and let h be a k-dimensional linear vector function on R^n. If

$$\left\langle \begin{array}{l} f(x) < 0 \\ h(x) = 0 \end{array} \right\rangle \text{ has no solution } x \in \Gamma$$

then there exist $p \in R^m$ and $q \in R^k$ such that

$$\left\langle \begin{array}{l} p \geq 0, \; (p,q) \neq 0 \\ pf(x) + qh(x) \geq 0 \; for \; all \; x \in \Gamma \end{array} \right\rangle$$

REMARK $p \geq 0$ and $(p,q) \neq 0$ does not imply $p \geq 0$ and $q \neq 0$, but it does imply $p \geq 0$ or $q \neq 0$ or both. However if we delete the linear equalities $h(x) = 0$, then $p \geq 0$.

PROOF Define the sets

$$\Lambda(x) = \{(y,z) \mid y \in R^m, z \in R^k, y > f(x), z = h(x)\} \qquad x \in \Gamma$$

and

$$\Lambda = \bigcup_{x \in \Gamma} \Lambda(x)$$

By hypothesis Λ does not contain the origin $0 \in R^{m+k}$. Also, Λ is convex, for if (y^1,z^1) and (y^2,z^2) are in Λ, then for $0 \leq \lambda \leq 1$

$$(1 - \lambda)y^1 + \lambda y^2 > (1 - \lambda)f(x^1) + \lambda f(x^2) \geq f[(1 - \lambda)x^1 + \lambda x^2]$$

and

$$(1 - \lambda)z^1 + \lambda z^2 = (1 - \lambda)h(x^1) + \lambda h(x^2) = h[(1 - \lambda)x^1 + \lambda x^2]$$

Because Λ is a nonempty convex set not containing the origin, it follows by Lemma 3.2.2 that there exist $p \in R^m$, $q \in R^k$, $(p,q) \neq 0$ such that

$$(u,v) \in \Lambda \Rightarrow pu + qv \geq 0$$

Since each u_i can be made as large as desired, $p \geq 0$.

Let $\epsilon > 0$, $u = f(x) + e\epsilon$, $v = h(x)$, $x \in \Gamma$, where e is a vector of ones in R^m. Hence $(u,v) \in \Lambda(x) \subset \Lambda$, and

$$pu + qv = pf(x) + \epsilon pe + qh(x) \geq 0 \qquad for \; x \in \Gamma$$

or

$$pf(x) + qh(x) \geq -\epsilon pe \qquad for \; x \in \Gamma$$

Now, if

$$\inf_{x \in \Gamma} pf(x) + qh(x) = -\delta < 0$$

we get, by picking ϵ such that $\epsilon pe < \delta$, that

$$\inf_{x \in \Gamma} pf(x) + qh(x) = -\delta < -\epsilon pe$$

which is a contradiction to the fact that $pf(x) + qh(x) \geq -\epsilon pe$ for all $x \in \Gamma$. Hence

$$\inf_{x \in \Gamma} pf(x) + qh(x) \geq 0 \quad \blacksquare$$

If we observe that for an m-dimensional vector function f defined on $\Gamma \subset R^n$ we have that

$$\left\langle \begin{array}{c} f(x) < 0 \\ \text{has a solution} \\ x \in \Gamma \end{array} \right\rangle \Rightarrow \left\langle \begin{array}{c} f(x) \leq 0 \\ \text{has a solution} \\ x \in \Gamma \end{array} \right\rangle \Rightarrow \left\langle \begin{array}{c} f(x) \leq 0 \\ \text{has a solution} \\ x \in \Gamma \end{array} \right\rangle$$

and

$$\left\langle \begin{array}{c} f(x) < 0 \\ \text{has no solution} \\ x \in \Gamma \end{array} \right\rangle \Leftarrow \left\langle \begin{array}{c} f(x) \leq 0 \\ \text{has no solution} \\ x \in \Gamma \end{array} \right\rangle \Leftarrow \left\langle \begin{array}{c} f(x) \leq 0 \\ \text{has no solution} \\ x \in \Gamma \end{array} \right\rangle$$

then the following corollary is a direct consequence of Theorem 1.

2   Corollary

Let Γ be a nonempty convex set in R^n, let f_1, f_2, f_3 be m^1-, m^2-, and m^3-dimensional convex vector functions on Γ, and h a k-dimensional linear vector function on R^n. If

$$\left\langle \begin{array}{c} f_1(x) < 0,\, f_2(x) \leq 0,\, f_3(x) \leq 0 \\ h(x) = 0 \end{array} \right\rangle \text{ has no solution } x \in \Gamma$$

then there exist $p_1 \in R^{m^1}$, $p_2 \in R^{m^2}$, $p_3 \in R^{m^3}$, and $q \in R^k$ such that

$$\left\langle \begin{array}{c} p_1,\, p_2,\, p_3 \geq 0,\, (p_1, p_2, p_3, q) \neq 0 \\ p_1 f_1(x) + p_2 f_2(x) + p_3 f_3(x) + qh(x) \geq 0 \quad \text{for all } x \in \Gamma \end{array} \right\rangle$$

We give now a generalization of Gordan's theorem of the alternative 2.4.5 to convex functions over an arbitrary convex set in R^n.

3   Generalized Gordan theorem [Fan-Glicksburg-Hoffman 57]

Let f be an m-dimensional convex vector function on the convex set $\Gamma \subset R^n$. Then either

I $f(x) < 0$ *has a solution* $x \in \Gamma$

or

II   *$pf(x) \geqq 0$ for all $x \in \Gamma$ for some $p \geq 0$, $p \in R^m$*

but never both.

PROOF   ($I \Rightarrow \overline{II}$)   Let $\bar{x} \in \Gamma$ be a solution of $f(x) < 0$.   Then for any $p \geq 0$ in R^m, $pf(\bar{x}) < 0$, and hence II cannot hold.
        ($\bar{I} \Rightarrow II$)   This follows directly from Theorem *1* above by deleting $h(x) = 0$ from the theorem.   ▌

    To see that *3* is indeed a generalization of Gordan's theorem *2.4.5* we let $f(x) = Ax$, where A is an $m \times n$ matrix.   Then

$$Ax > 0 \text{ has no solution } x \in R^n \Leftrightarrow \left\langle \begin{array}{l} pAx \geqq 0 \quad \text{for all } x \in R^n \\ \text{for some } p \geq 0, \, p \in R^m \quad \text{(by } 3) \end{array} \right\rangle$$

$$\Leftrightarrow A'p = 0, \, p \geq 0 \quad \text{for some } p \in R^m$$

where the last equivalence follows by taking $x = \pm e^i$, $i = 1, \ldots, n$, where $e^i \in R^n$ has zeros for all its elements except 1 for the ith element.
    In the same spirit, Theorem *1* above can be considered a partial generalization of Motzkin's theorem *2.4.2* to convex functions.   The generalization is partial (unlike *3* which is a complete generalization of Gordan's theorem), because the statement of Theorem *1* does not exclude the possibility of both systems having a solution, that is, there may exist an $\bar{x} \in \Gamma$ and $p \geqq 0$, $(p,q) \neq 0$, such that $f(\bar{x}) < 0$, $h(\bar{x}) = 0$, and $pf(x) + qh(x) \geqq 0$ for all $x \in \Gamma$.   Similarly, Corollary *2* is a partial generalization of Slater's theorem *2.4.1*.   However, it is possible to sharpen Theorem *1* and make it a theorem of the alternative if we let $\Gamma = R^n$, $h(x) = Bx - d$ and require that the rows of B be linearly independent.   We obtain then the following result.

4   **Theorem**

    Let f be a given m-dimensional convex function on R^n, let B be a given $k \times n$ matrix with linearly independent rows, and let d be a given k-dimensional vector.   Then either

I   *$f(x) < 0$, $Bx = d$ has a solution $x \in R^n$*

or

II   *$pf(x) + q(Bx - d) \geqq 0$ for all $x \in R^n$ for some $p \geq 0$, $p \in R^m$,*
$$q \in R^k$$

but never both.

PROOF (I $\Rightarrow \overline{\text{II}}$) Let $\bar{x} \in R^n$ be a solution of $f(x) < 0$ and $Bx = d$. Then for any $p \geq 0$ and q in R^m and R^k respectively,

$$pf(\bar{x}) + q(B\bar{x} - d) < 0$$

Hence II cannot hold.

($\bar{\text{I}} \Rightarrow$ II) If I has no solution then by Theorem *1* there exists $p \geq 0$, $(p,q) \neq 0$ such that

$$pf(x) + q(Bx - d) \geq 0 \qquad \text{for all } x \in R^n$$

If $p \geq 0$, the theorem is proved. We assume the contrary, that $p = 0$, and exhibit a contradiction. If $p = 0$, then

$$q(Bx - d) \geq 0 \qquad \text{for all } x \in R^n \text{ for some } q \neq 0$$

We will show now that $B'q = 0$. For, if $B'q \neq 0$, then by picking $x = -qB$ for the case when $qd \geq 0$, and $x = 2(qd)qB/qBB'q$ for the case when $qd < 0$, we obtain that $q(Bx - d) < 0$. Hence $B'q = 0$ for some $q \neq 0$, which contradicts the assumption that the rows of B are linearly independent. ∎

We close this chapter by obtaining another fundamental theorem for a (possibly infinite) family of convex and linear functions.

5   ## Theorem [Bohnenblust-Karlin-Shapley 50]

Let Γ be a nonempty compact convex set in R^n and let $(f_i)_{i \in M}$ be a family (finite or infinite) of numerical functions which are convex and lower semicontinuous on Γ, and let $(h_i)_{i \in K}$ be a family (finite or infinite) of linear numerical functions on R^n. If

$$\left\langle \begin{array}{l} f_i(x) \leq 0, \, i \in M \\ h_i(x) = 0, \, i \in K \end{array} \right\rangle \quad \textit{has no solution } x \in \Gamma$$

then for some finite subfamily $(f_{i_1}, \ldots, f_{i_m})$ of $(f_i)_{i \in M}$ and some finite subfamily $(h_{i_1}, \ldots, h_{i_k})$ of $(h_i)_{i \in K}$ there exist $p \in R^m$ and $q \in R^k$ such that

$$\left\langle \begin{array}{l} p \geq 0, \, (p,q) \neq 0 \\ \displaystyle\sum_{j=1}^{m} p_j f_{i_j}(x) + \sum_{j=1}^{k} q_j h_{i_j}(x) \geq 0 \qquad \text{for all } x \in \Gamma \end{array} \right\rangle$$

If K is empty, that is if all equalities $h_i(x) = 0$ are deleted, then the last inequality above (≥ 0) becomes a strict inequality (> 0).

PROOF   [Berge–Ghouila Houri 65]   The system

$$\left\langle \begin{array}{l} f_i(x) \leqq \epsilon, \ \forall i \in M, \ \forall \epsilon > 0 \\ h_i(x) = 0, \ \forall i \in K \end{array} \right\rangle$$

has no solution x in Γ. [For if it did have a solution $\bar{x}$, then $f_i(\bar{x}) \leqq \epsilon$ for all $\epsilon > 0$ and all $i \in M$, and $h_i(\bar{x}) = 0$ for all $i \in K$. This in turn implies that $f_i(\bar{x}) \leqq 0$ for all $i \in M$ and $h_i(\bar{x}) = 0$ for all $i \in K$ (for otherwise if $f_i(\bar{x}) > 0$ for some $i \in M$, then picking $\epsilon = \frac{1}{2}f_i(\bar{x}) > 0$ would lead to a contradiction). This however contradicts the hypothesis of the theorem.] The sets

$$\Gamma(i,j,\epsilon) = \{x \mid x \in \Gamma, f_i(x) \leqq \epsilon, h_j(x) = 0\}$$

are closed sets (because of the lower semicontinuity of f_i, the linearity of h_j, and the compactness of Γ, see Appendix C) contained in the compact set Γ, and their intersection is empty. Hence by the finite intersection theorem $B.3.2$(iii) there exist a finite number of such sets so that their intersection is empty. Thus we obtain indices $(i_1, i_2, \ldots, i_m) \in M$, $(i_1, i_2, \ldots, i_k) \in K$, and real numbers $\epsilon_1, \epsilon_2, \ldots, \epsilon_m > 0$, such that the system

$$\left\langle \begin{array}{l} f_{i_j}(x) - \epsilon_j \leqq 0, \ j = 1, \ldots, m \\ h_{i_j}(x) = 0, \ j = 1, \ldots, k \end{array} \right\rangle$$

has no solution $x \in \Gamma$. Hence by Corollary 2 there exist $p \in R^m, q \in R^k$ such that

$$p \geqq 0 \qquad (p,q) \neq 0$$

and

$$\sum_{j=1}^{m} p_j f_{i_j}(x) + \sum_{j=1}^{k} q_j h_{i_j}(x) \geqq \sum_{j=1}^{m} p_j \epsilon_j \qquad \text{for all } x \in \Gamma$$

from which the conclusion of the theorem follows if we observe that $\sum_{j=1}^{m} p_j \epsilon_j \geqq 0$, and if K is empty, then $p \geq 0$ and $\sum_{j=1}^{m} p_j \epsilon_j > 0$. ∎

Chapter Five

Saddlepoint Optimality Criteria of Nonlinear Programming Without Differentiability

The purpose of this chapter is to derive optimality criteria of the saddlepoint type for nonlinear programming problems. This type of optimality criterion is perhaps best illustrated by a simple example. Consider the problem of minimizing the function θ on the set $X = \{x \mid x \in R, \ -x + 2 \leq 0\}$, where $\theta(x) = (x)^2$. Obviously the solution is $\bar{x} = 2$, and the minimum is $\theta(\bar{x}) = 4$. The saddlepoint optimality criterion for this problem is this: A necessary and sufficient condition that $\bar{x}$ be a solution of the minimization problem is that there exists a real number $\bar{u}$ (here $\bar{u} = 4$) such that for all $x \in R$ and all $u \in R, \ u \geq 0$

$$
\begin{aligned}
\theta(\bar{x}) &+ u(-\bar{x} + 2) \\
&\leq \theta(\bar{x}) + \bar{u}(-\bar{x} + 2) \\
&\qquad \leq \theta(x) + \bar{u}(-x + 2)
\end{aligned}
$$

It is easy to verify that the above inequalities are satisfied for $\bar{x} = 2$, $\bar{u} = 4$. Hence the function ψ defined on R^2 by

$$
\psi(x,u) = \theta(x) + u(-x + 2)
$$

has a saddlepoint at $\bar{x} = 2$, $\bar{u} = 4$, because it has a minimum at $(\bar{x},\bar{u})$ with respect to x for all real x, and a maximum with respect to u for all real nonnegative u.

For the above simple problem, the saddlepoint criterion happens to be both a necessary and a sufficient optimality criterion for $\bar{x}$ to be a solution of the minimization problem. This is not always the case. We shall show in this chapter that the above saddlepoint condition is a sufficient optimality condition without any convexity requirements. However to establish the necessity of the above saddle-point condition, we need not only convexity but also some sort of a regularity condition, a constraint qualification. This confirms earlier statements made to the effect that necessary optimality conditions are more complex and harder to establish.

We shall develop the optimality criteria of this chapter without any differentiability assumptions on the functions involved. Subsequent chapters, Chaps. 7 and 11, will establish optimality criteria that involve differentiable functions.

1. The minimization and saddlepoint problems

The optimality criteria of this chapter relate the solutions of a minimization problem, a local minimization problem, and two saddlepoint problems to each other. We define these problems below now.

Let X^0 be a subset of R^n, let θ and g be respectively a numerical function and an m-dimensional vector function defined on X^0.

1   **The minimization problem (MP)**

Find an $\bar{x}$, if it exists, such that

$$\theta(\bar{x}) = \min_{x \in X} \theta(x) \qquad \bar{x} \in X = \{x \mid x \in X^0, g(x) \leq 0\} \qquad \text{(MP)}$$

The set X is called the *feasible region* or the *constraint set*, $\bar{x}$ the *minimum solution* or *solution*, and $\theta(\bar{x})$ the *minimum*. All points x in the feasible region X are called *feasible* points.

If X is a convex set, and if θ is convex on X, the minimization problem MP is often called a *convex programming problem* or *convex program*.

(We observe that the above minimization problem is a special case of the general minimization problem *1.6.9*, where the additional k-dimensional vector equality constraint $h(x) = 0$ was also present. The reason for this is that in the absence of differentiability there are no significant optimality criteria for problems with nonlinear equality constraints. Some results for linear equality constraints will be obtained however. See *5.3.2*, *5.4.2*, and *5.4.8*.)

2   **The local minimization problem (LMP)**

Find an $\bar{x}$ in X, if it exists, such that for some open ball $B_\delta(\bar{x})$ around $\bar{x}$ with radius $\delta > 0$

$$x \in B_\delta(\bar{x}) \cap X \Rightarrow \theta(x) \geq \theta(\bar{x}) \qquad \text{(LMP)}$$

3   **The Fritz John saddlepoint problem (FJSP)**

Find $\bar{x} \in X^0$, $\bar{r}_0 \in R$, $\bar{r} \in R^m$, $(\bar{r}_0, \bar{r}) \geq 0$, if they exist, such that

$$\left/ \begin{aligned} &\phi(\bar{x}, \bar{r}_0, r) \leq \phi(\bar{x}, \bar{r}_0, \bar{r}) \leq \phi(x, \bar{r}_0, \bar{r}) \\ &\text{for all } r \geq 0,\ r \in R^m,\ \text{and all } x \in X^0, \\ &\phi(x, r_0, r) = r_0 \theta(x) + r g(x) \end{aligned} \right\rangle \qquad \text{(FJSP)}$$

4   **The Kuhn-Tucker saddlepoint problem (KTSP)**

Find $\bar{x} \in X^0$, $\bar{u} \in R^m$, $\bar{u} \geq 0$, if they exist, such that

$$\left/ \begin{aligned} &\psi(\bar{x}, u) \leq \psi(\bar{x}, \bar{u}) \leq \psi(x, \bar{u}) \\ &\text{for all } u \geq 0,\ u \in R^m,\ \text{and all } x \in X^0, \\ &\psi(x, u) = \theta(x) + u g(x) \end{aligned} \right\rangle \qquad \text{(KTSP)}$$

5   **Remark**

If $(\bar{x}, \bar{r}_0, \bar{r})$ is a solution of FJSP and $\bar{r}_0 > 0$, then $(\bar{x}, \bar{r}/\bar{r}_0)$ is a solution of KTSP. Conversely, if $(\bar{x}, \bar{u})$ is a solution of KTSP, then $(\bar{x}, 1, \bar{u})$ is a solution of FJSP.

6   **Remark**

The numerical functions $\phi(x, r_0, r)$ and $\psi(x, u)$ defined above are often called *Lagrangian functions* or simply *Lagrangians*, and the m-dimensional vectors $\bar{r}$ and $\bar{u}$ *Lagrange multipliers* or *dual variables*. These multipliers play a role in linear and nonlinear programming which is very similar to the role played by the Lagrange multipliers of the classical calculus where a function of several variables is to be minimized subject to equality constraints (see for example [Fleming 65]). Here, because we have inequality constraints, the Lagrange multipliers turn out to be nonnegative. When we shall consider equality constraints in *5.3.2*, *5.4.2*, and *5.4.8*, the multiplier associated with these equalities will not be required to be nonnegative.

7   **Remark**

The right inequality of both saddlepoint problems, FJSP *3* and KTSP *4*

$$\phi(\bar{x}, \bar{r}_0, \bar{r}) \leq \phi(x, \bar{r}_0, \bar{r}) \qquad \text{for all } x \in X^0$$

and

$$\psi(\bar{x}, \bar{u}) \leq \psi(x, \bar{u}) \qquad \text{for all } x \in X^0$$

can be interpreted as a *minimum principle*, akin to *Pontryagin's maximum principle*† [Pontryagin et al. 62]. Pontryagin's principle in its original form is a necessary optimality condition for the optimal control of systems described by ordinary differential equations. As such, it is a necessary optimality condition for a programming problem, not in R^n, but in some other space. More recently [Halkin 66, Canon et al. 66, Mangasarian-Fromovitz 67] a minimum principle has also been established for optimal control problems described by ordinary difference equations. This is a programming problem in R^n, which unfortunately is not convex in general, and hence the results of this chapter do not apply. However the optimality conditions of Chaps. 7 and 11, which are based mainly on linearization and not on convexity, do apply to optimal control problems described by nonlinear difference equations.

2.   Some basic results for minimization and local minimization problems

We establish now some basic results concerning the set of solutions of the minimization problem and relate the solutions of the minimization and local minimization problems to each other.

1   **Theorem**

Let X be a convex set, and let θ be a convex function on X. The set of solutions of MP 5.1.1 is convex.

REMARK   A sufficient but not necessary condition for the convexity of X is that X^0 be a convex set and that g be convex on X^0. This follows from *4.1.10* and *3.1.9*.

PROOF   Let x^1 and x^2 be solutions of MP. That is,

$$\theta(x^1) = \theta(x^2) = \min_{x \in X} \theta(x)$$

It follows by the convexity of X and θ, that for $0 \leq \lambda \leq 1$, $(1 - \lambda)x^1 + \lambda x^2 \in X$, and

$$\theta[(1 - \lambda)x^1 + \lambda x^2] \leq (1 - \lambda)\theta(x^1) + \lambda\theta(x^2) = \theta(x^1) = \min_{x \in X} \theta(x)$$

Hence $(1 - \lambda)x^1 + \lambda x^2$ is also a solution of MP, and the set of solutions is convex. ∎

† Pontryagin gets a maximum principle instead of a minimum principle because his Lagrangian is the negative of the Lagrangian of nonlinear programming.

2 **Uniqueness theorem**

Let X be convex and $\bar{x}$ be a solution of MP 5.1.1. If θ is strictly convex at $\bar{x}$, then $\bar{x}$ is the unique solution of MP.

PROOF Let $\hat{x} \neq \bar{x}$ be another solution of MP, that is, $\hat{x} \in X$, and $\theta(\hat{x}) = \theta(\bar{x})$. Since X is convex, then $(1 - \lambda)\bar{x} + \lambda\hat{x} \in X$ whenever $0 < \lambda < 1$, and by the strict convexity of θ at $\bar{x}$

$$\theta[(1 - \lambda)\bar{x} + \lambda\hat{x}] < (1 - \lambda)\theta(\bar{x}) + \lambda\theta(\hat{x}) = \theta(\bar{x})$$

This contradicts the assumption that $\theta(\bar{x})$ is a minimum, and hence $\hat{x}$ cannot be another solution.

3 **Theorem**

Let X be convex, and let θ be a nonconstant concave function on X. Then no interior point of X is a solution of MP 5.1.1, or equivalently any solution $\bar{x}$ of MP, if it exists, must be a boundary point of X.

PROOF If MP *5.1.1* has no solution the theorem is trivially true. Let $\bar{x}$ be a solution of MP. Since θ is not constant on X, there exists a point $x \in X$ such that $\theta(x) > \theta(\bar{x})$. If z is an interior point of X, there exists a point $y \in X$ such that for some λ, $0 \leqq \lambda < 1$

$$z = (1 - \lambda)x + \lambda y$$

See Fig. *5.2.1*. Hence

$$\theta(z) = \theta[(1 - \lambda)x + \lambda y] \geqq (1 - \lambda)\theta(x) + \lambda\theta(y)$$
$$> (1 - \lambda)\theta(\bar{x}) + \lambda\theta(\bar{x})$$
$$= \theta(\bar{x})$$

and $\theta(x)$ does not attain its minimum at an interior point z. ∎
Figure *5.2.2* shows a simple example of Theorem *3* in R.

4 **Theorem**

If $\bar{x}$ is a solution of MP 5.1.1, then it is also a solution of LMP 5.1.2. The converse is true if X is convex and θ is convex at $\bar{x}$.

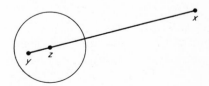

Fig. 5.2.1

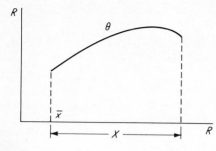

Fig. 5.2.2 A simple example of Theorem
3 in R.

PROOF If $\bar{x}$ solves MP, then $\bar{x}$ solves LMP for any $\delta > 0$. To prove
the converse now, assume that $\bar{x}$ solves LMP for some $\delta > 0$, and let X
be convex and θ be convex at $\bar{x}$. Let y be any point in X distinct from $\bar{x}$.
Since X is convex, $(1 - \lambda)\bar{x} + \lambda y \in X$ for $0 < \lambda \leq 1$. By choosing λ
small enough, that is, $0 < \lambda < \delta/\|y - \bar{x}\|$ and $\lambda \leq 1$, we have that

$$\bar{x} + \lambda(y - \bar{x}) = (1 - \lambda)\bar{x} + \lambda y \in B_\delta(\bar{x}) \cap X$$

Hence

$$\theta(\bar{x}) \leq \theta[\bar{x} + \lambda(y - \bar{x})] \qquad \text{(since } \bar{x} \text{ solves LMP)}$$

$$\leq (1 - \lambda)\theta(\bar{x}) + \lambda\theta(y) \qquad \text{(by convexity of } \theta \text{ at } \bar{x})$$

from which it follows that

$$\theta(\bar{x}) \leq \theta(y) \quad \blacksquare$$

3. Sufficient optimality criteria

The main sufficient optimality criteria developed here (*1* and *2* below)
require no convexity assumptions on the minimization problem MP *5.1.1*.
These criteria are quite straightforward to obtain and need no complicated machinery to derive. First results of this type were obtained in
[Uzawa 58].

1   **Sufficient optimality theorem**

*If $(\bar{x},\bar{u})$ is a solution of KTSP 5.1.4, then $\bar{x}$ is a solution of MP 5.1.1.
If $(\bar{x},\bar{r}_0,\bar{r})$ is a solution of FJSP 5.1.3, and $\bar{r}_0 > 0$, then $\bar{x}$ is a solution of
MP 5.1.1.*

PROOF The second statement of the theorem follows trivially from the
first statement by Remark *5.1.5*.
Let $(\bar{x},\bar{u})$ be a solution of KTSP *5.1.4*. Then for all $u \geq 0$ in R^m

and all x in X^0

$$\theta(\bar{x}) + ug(\bar{x}) \leqq \theta(\bar{x}) + \bar{u}g(\bar{x}) \leqq \theta(x) + \bar{u}g(x)$$

From the first inequality we have that

$$(u - \bar{u})g(\bar{x}) \leqq 0 \qquad \text{for all } u \geqq 0$$

For any j, $1 \leqq j \leqq m$, let

$$\left\langle \begin{array}{ll} u_i = \bar{u}_i & \text{for } i = 1, 2, \ldots, j-1, j+1, \ldots, m \\ u_j = \bar{u}_j + 1 \end{array} \right\rangle$$

It follows then that $g_j(\bar{x}) \leqq 0$. Repeating this for all j, we get that $g(\bar{x}) \leqq 0$, and hence $\bar{x}$ is a feasible point, that is, $\bar{x} \in X$.

Now since $\bar{u} \geqq 0$ and $g(\bar{x}) \leqq 0$, we have that $\bar{u}g(\bar{x}) \leqq 0$. But again from the first inequality of the saddlepoint problem we have, by setting $u = 0$, that $\bar{u}g(\bar{x}) \geqq 0$. Hence $\bar{u}g(\bar{x}) = 0$.

Let x be any point in X, then from the second inequality of the saddlepoint problem we get

$$\theta(\bar{x}) \leqq \theta(x) + \bar{u}g(x) \qquad [\text{since } \bar{u}g(\bar{x}) = 0]$$
$$\leqq \theta(x) \qquad\qquad [\text{since } \bar{u} \geqq 0, g(x) \leqq 0]$$

Hence $\bar{x}$ is a solution of MP. ∎

It should be remarked here that because no convexity assumptions were made in the above theorem, equality constraints can be handled by replacing them by two inequality constraints. That is, replace $h(x) = 0$ by $h(x) \leqq 0$ and $-h(x) \leqq 0$.

2   **Problem**

Consider the minimization problem

$$\theta(\bar{x}) = \min_{x \in X} \theta(x) \qquad \bar{x} \in X = \{x \mid x \in X^0, g(x) \leqq 0, h(x) = 0\}$$

where h is a k-dimensional vector function on X^0 and all else is defined as in MP *5.1.1*. Let

$$\phi(x,r_0,r,s) = r_0\theta(x) + rg(x) + sh(x)$$

and

$$\psi(x,u,v) = \theta(x) + ug(x) + vh(x)$$

Show that if there exist $\bar{x} \in X^0$, $\bar{u} \in R^m$, $\bar{u} \geqq 0$, $\bar{v} \in R^k$ such that

$$\left\langle \begin{array}{l} \psi(\bar{x},u,v) \leqq \psi(\bar{x},\bar{u},\bar{v}) \leqq \psi(x,\bar{u},\bar{v}) \\ \text{for all } u \geqq 0, u \in R^m, \text{ all } v \in R^k, \text{ and all } x \in X^c \end{array} \right\rangle$$

or if there exist $\bar{x} \in X^0$, $\bar{r}_0 \in R$, $\bar{r}_0 > 0$, $\bar{r} \in R^m$, $\bar{r} \geqq 0$, $\bar{s} \in R^k$ such that

$$\left\langle \begin{array}{l} \phi(\bar{x},\bar{r}_0,r,s) \leqq \phi(\bar{x},\bar{r}_0,\bar{r},\bar{s}) \leqq \phi(x,\bar{r}_0,\bar{r},\bar{s}) \\ \text{for all } r \geqq 0, \; r \in R^m, \text{ all } s \in R^k, \text{ and all } x \in X^0 \end{array} \right\rangle$$

then $\bar{x}$ is a solution of the minimization problem. (Notice that v and s are not restricted in sign.)

The question may be raised as to what sort of point is the point $\bar{x}$ if $(\bar{x},\bar{r}_0,\bar{r})$ is a solution of FJSP *5.1.3* and we do not require that $\bar{r}_0 > 0$. An answer to this question is given by the following result.

3

Corollary

If $(\bar{x},\bar{r}_0,\bar{r})$ is a solution of FJSP 5.1.3, then either $\bar{x}$ solves MP 5.1.1 or X has no interior relative to $g(x) \leqq 0$, that is, $\{x \mid x \in X^0, g(x) < 0\} = \emptyset$.

PROOF By the same argument as in the proof of Theorem *1* above we show that $g(\bar{x}) \leqq 0$ and $\bar{r}g(\bar{x}) = 0$. Now, if $\bar{r}_0 > 0$, then $\bar{x}$ solves MP by Theorem *1*. If $\bar{r}_0 = 0$, then $\bar{r} \geq 0$ and we have from the second inequality of FJSP *5.1.3* that

$$0 = \bar{r}g(\bar{x}) \leqq \bar{r}g(x) \qquad \text{for all } x \in X^0$$

Now, if the set $\{x \mid x \in X^0, g(x) < 0\}$ is nonempty, then for any element $\hat{x}$ in it $\bar{r}g(\hat{x}) < 0$, which contradicts the fact established above that $\bar{r}g(x) \geqq 0$ for all $x \in X^0$. Hence $\{x \mid x \in X^0, g(x) < 0\} = \emptyset$. ∎

4. Necessary optimality criteria

The situation with respect to necessary criteria is considerably more complicated than the situation with respect to sufficient optimality criteria. The two situations are compared in the following table:

Necessary criteria	Sufficient criteria
(a) Convexity needed	No convexity needed
(b) Consequence of separation theorem of convex sets needed	Separation theorem of convex sets not needed
(c) Regularity condition (constraint qualification) needed in the more important necessary criterion (7 below)	No constraint qualification needed

We begin by establishing a necessary optimality criterion which does not require any regularity conditions. This necessary optimality criterion is similar in spirit to the necessary optimality criterion of Fritz John [John 48] (see also Chap. 7), which was derived for the case where the functions θ and g were differentiable but not convex. We use no differentiability here, but instead we use convexity. The present criterion is a saddle-point criterion, whereas Fritz John's is a gradient criterion. The main point of similarity is the presence of the multiplier $\bar{r}_0$ in both criteria.

1   **Fritz John saddlepoint necessary optimality theorem**
         [Uzawa 58, Karlin 59]

Let X^0 be a convex set in R^n, and let θ and g be convex on X^0. If $\bar{x}$ is a solution of MP 5.1.1, then $\bar{x}$ and some $\bar{r}_0 \in R$, $\bar{r} \in R^m$, $(\bar{r}_0, \bar{r}) \geq 0$ solve FJSP 5.1.3 and $\bar{r}g(\bar{x}) = 0$.

PROOF   Because $\bar{x}$ solves MP

$$\left\langle \begin{array}{c} \theta(x) - \theta(\bar{x}) < 0 \\ g(x) \leq 0 \end{array} \right\rangle \quad \text{has no solution } x \in X^0$$

By Corollary 4.2.2 there exist $\bar{r}_0 \in R$, $\bar{r} \in R^m$, $(\bar{r}_0, \bar{r}) \geq 0$ such that

$$\bar{r}_0[\theta(x) - \theta(\bar{x})] + \bar{r}g(x) \geq 0 \qquad \text{for all } x \in X^0$$

By letting $x = \bar{x}$ in the above, we get that $\bar{r}g(\bar{x}) \geq 0$. But since $\bar{r} \geq 0$ and $g(\bar{x}) \leq 0$, we also have $\bar{r}g(\bar{x}) \leq 0$. Hence

$$\bar{r}g(\bar{x}) = 0$$

and

$$\bar{r}_0\theta(\bar{x}) + \bar{r}g(\bar{x}) \leq \bar{r}_0\theta(x) + \bar{r}g(x) \qquad \text{for all } x \in X^0$$

which is the second inequality of FJSP *5.1.3*.
     We also have, because $g(\bar{x}) \leq 0$, that

$$rg(\bar{x}) \leq 0 \qquad \text{for all } r \geq 0, \, r \in R^m$$

and hence, since $\bar{r}g(\bar{x}) = 0$

$$\bar{r}_0\theta(\bar{x}) + rg(\bar{x}) \leq \bar{r}_0\theta(\bar{x}) + \bar{r}g(\bar{x}) \qquad \text{for all } r \geq 0, \, r \in R^m$$

which is the first inequality of FJSP *5.1.3*.   ∎

2   **Problem**
     Consider the minimization problem

$$\theta(\bar{x}) = \min_{x \in X} \theta(x) \qquad \bar{x} \in X = \{x \mid x \in X^0, g(x) \leq 0, h(x) = 0\}$$

where h is a k-dimensional linear vector function on R^n, θ and g are convex on X^0, and all else is defined as in MP $5.1.1$. Show that if $\bar{x}$ is a solution of the above problem, then $\bar{x}$ and some $\bar{r}_0 \in R$, $\bar{r} \in R^m$, $\bar{s} \in R^k$, $(\bar{r}_0, \bar{r}) \geqq 0$, $(\bar{r}_0, \bar{r}, \bar{s}) \neq 0$ satisfy $\bar{r}g(\bar{x}) = 0$, and

$$\left\langle \begin{array}{l} \phi(\bar{x}, \bar{r}_0, r, s) \leqq \phi(\bar{x}, \bar{r}_0, \bar{r}, \bar{s}) \leqq \phi(x, \bar{r}_0, \bar{r}, \bar{s}) \\[2mm] \quad \text{for all } r \geqq 0,\ r \in R^m,\ \text{all } s \in R^k,\ \text{and all } x \in X^0 \\[2mm] \phi(x, r_0, r, s) = r_0\theta(x) + rg(x) + sh(x) \end{array} \right\rangle$$

(Hint: Again use Corollary $4.2.2$.)

It should be remarked here that in the above necessary optimality criteria there is no guarantee that $\bar{r}_0 > 0$. In cases where $\bar{r}_0 = 0$ it is intuitively obvious that the necessary optimality criterion FJSP $5.1.3$ does not say much about the minimization problem MP $5.1.1$, because the function θ has disappeared from $5.1.3$ and any other function could have played its role. In order to exclude such cases, we have to introduce some regularity conditions. These regularity conditions are referred to in the literature as *constraint qualifications*. We shall have occasion to use a number of these constraint qualifications throughout this book. Some of these constraint qualifications (like the three introduced below) make use only of the convexity properties of the functions defining the feasible region X. Other constraint qualifications, to be introduced later, in Chap. 7 for example, make use mostly of the differentiability properties of the functions defining the feasible region X.

3   Slater's constraint qualification [Slater 50]

Let X^0 be a convex set in R^n. The m-dimensional convex vector function g on X^0 which defines the convex feasible region

$$X = \{x \mid x \in X^0,\ g(x) \leqq 0\}$$

is said to satisfy *Slater's constraint qualification (on X^0)* if there exists an $\bar{x} \in X^0$ such that $g(\bar{x}) < 0$.

4   Karlin's constraint qualification [Karlin 59]

Let X^0 be a convex set in R^n. The m-dimensional convex vector function g on X^0 which defines the convex feasible region

$$X = \{x \mid x \in X^0,\ g(x) \leqq 0\}$$

is said to satisfy *Karlin's constraint qualification (on X^0)* if there exists *no* $p \in R^m$, $p \geq 0$ such that

$$pg(x) \geqq 0 \qquad \text{for all } x \in X^0$$

5   The strict constraint qualification

Let X^0 be a convex set in R^n. The m-dimensional convex vector function g on X^0 which defines the convex feasible region

$$X = \{x \mid x \in X^0,\, g(x) \leq 0\}$$

is said to satisfy the *strict constraint qualification* (*on* X^0) if X contains at least two distinct points x^1 and x^2 such that g is strictly convex at x^1.

6   **Lemma**

Slater's constraint qualification 3 and Karlin's constraint qualification 4 are equivalent. The strict constraint qualification 5 implies Slater's and Karlin's constraint qualifications 3 and 4.

PROOF $(3 \Leftrightarrow 4)$ By Gordan's generalized theorem 4.2.3, 3 and 4 are equivalent.

$(5 \Rightarrow 3)$ Since X^0 is convex, for any λ, $0 < \lambda < 1$

$$(1 - \lambda)x^1 + \lambda x^2 \in X^0$$

Because g is strictly convex at x^1, it follows from 4.1.4 that

$$g[(1 - \lambda)x^1 + \lambda x^2)] < (1 - \lambda)g(x^1) + \lambda g(x^2) \leq 0$$

where the last inequality follows from the fact that $g(x^1) \leq 0$ and $g(x^2) \leq 0$. Thus g satisfies Slater's constraint qualification 3 and hence also Karlin's. ∎

We are ready now to derive the most important necessary optimality criterion without the use of differentiability. The theorem is widely known under the name Kuhn-Tucker [Kuhn-Tucker 51], even though Kuhn and Tucker required both convexity and differentiability in its derivation. The theorem in its present form, without any differentiability requirements, is attributed to Uzawa [Uzawa 58] and Karlin [Karlin 59].

7   **Kuhn-Tucker saddlepoint necessary optimality theorem**
          [Kuhn-Tucker 51, Uzawa 58, Karlin 59]

Let X^0 be a convex set in R^n, let θ and g be convex on X^0, and let g satisfy Slater's constraint qualification 3, Karlin's constraint qualification 4, or the strict constraint qualification 5 on X^0. If $\bar{x}$ is a solution of MP 5.1.1, then $\bar{x}$ and some $\bar{u} \in R^m$, $\bar{u} \geq 0$, solve KTSP 5.1.4 and $\bar{u}g(\bar{x}) = 0$.

PROOF   We first observe that by Lemma 6 above we need only establish the theorem under Karlin's constraint qualification. By Theorem 1, $\bar{x}$ and some $\bar{r}_0 \in R$, $\bar{r} \in R^m$, $(\bar{r}_0, \bar{r}) \geq 0$, solve FJSP $5.1.3$ and $\bar{r}g(\bar{x}) = 0$. If $\bar{r}_0 > 0$, then by Remark $5.1.5$ we are done. If $\bar{r}_0 = 0$, then $\bar{r} \geq 0$, and from the second inequality of FJSP $5.1.3$

$$0 \leq \bar{r}g(x) \qquad \text{for all } x \in X^0 \qquad [\text{since } \bar{r}_0 = 0 \text{ and } \bar{r}g(\bar{x}) = 0]$$

which contradicts Karlin's constraint qualification 4. Hence $\bar{r}_0 > 0$. ∎

We summarize in Fig. $5.4.1$ the relationships between the solutions of the various problems of this chapter.

We end this section by deriving a Kuhn-Tucker saddlepoint necessary optimality criterion in the presence of linear equality constraints. In order to do this, we have to let the set X^0 of MP $5.1.1$ be the entire space R^n.

8   **Kuhn-Tucker saddlepoint necessary optimality theorem in the presence of linear equality constraints [Uzawa 58]**

Let θ, g be respectively a numerical function and an m-dimensional vector function which are both convex on R^n. Let h be a k-dimensional linear vector function on R^n, that is, $h(x) = Bx - d$, where B is a $k \times n$ matrix, and d is a k-vector. Let $\bar{x}$ be a solution of the minimization problem

$$\theta(\bar{x}) = \min_{x \in X} \theta(x) \qquad \bar{x} \in X = \{x \mid x \in R^n, g(x) \leq 0, Bx = d\}$$

and let g and h satisfy any of the constraint qualifications:

(i)   *(Generalized Slater 3) $g(x) < 0$, $Bx = d$ has a solution $x \in R^n$*

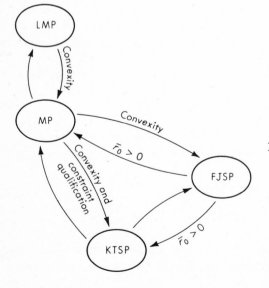

Fig. 5.4.1 Relationships between the solutions of the local minimization problem (LMP) $5.1.2$, the minimization problem (MP) $5.1.1$, the Fritz John saddlepoint problem (FJSP) $5.1.3$, and the Kuhn-Tucker saddlepoint problem (KTSP) $5.1.4$.

(ii)   (*Generalized Karlin 4*)   *There exists no* $p \geq 0$, $p \in R^m$, $q \in R^k$
        such that

$$pg(x) + q(Bx - d) \geq 0 \qquad for\ all\ x \in R^n$$

(iii)   (*Generalized strict 5*) X *contains at least two distinct points* x^1 *and* x^2
        such that g *is strictly convex at* x^1

Then $\bar{x}$ *and some* $\bar{u} \in R^m$, $\bar{u} \geq 0$, $\bar{v} \in R^k$ *satisfy* $\bar{u}g(\bar{x}) = 0$, *and*

$$\left/ \begin{array}{l} \phi(\bar{x},u,v) \leq \phi(\bar{x},\bar{u},\bar{v}) \leq \phi(x,\bar{u},\bar{v}) \\[4pt] for\ all\ u \geq 0,\ u \in R^m,\ all\ v \in R^k,\ and\ all\ x \in R^n \\[4pt] \phi(x,u,v) = \theta(x) + ug(x) + v(Bx - d) \end{array} \right\rangle \right.$$

PROOF   We shall first establish the fact that

(iii) $\Rightarrow$ (i) $\Rightarrow$ (ii)

and then prove the theorem under (ii).
        [(iii) $\Rightarrow$ (i)]   Since $g(x^1) \leq 0$, $g(x^2) \leq 0$, $Bx^1 = d$, $Bx^2 = d$, we
have for $0 < \lambda < 1$ that $B[(1 - \lambda)x^1 + \lambda x^2] = d$ and

$$g[(1 - \lambda)x^1 + \lambda x^2] < (1 - \lambda)g(x^1) + \lambda g(x^2) \leq 0$$

Hence (i) holds.
        [(i) $\Rightarrow$ (ii)]   If $g(\bar{x}) < 0$ and $B\bar{x} = d$, then for any $p \geq 0$, $p \in r^m$,
and any $q \in R^k$,

$$pg(\bar{x}) + q(B\bar{x} - d) < 0$$

Hence (ii) holds.
        We establish now the theorem under (ii).   There will be no loss
of generality if we assume that the rows $B_1, \ldots, B_k$ of B are linearly
independent, for suppose that some row, B_k say, is linearly dependent
on $B_1, \ldots, B_{k-1}$, that is $B_k = \sum\limits_{i=1}^{k-1} s_iB_i$, where $s_1, \ldots, s_{k-1}$ are fixed
real numbers.   Then

$$B_kx - d_k = \sum_{i=1}^{k-1} s_iB_ix - d_k = \sum_{i=1}^{k-1} s_id_i - d_k$$

for any x satisfying $B_ix = d_i$, $i = 1, \ldots, k - 1$.   But since $\bar{x} \in X$,
and $B_i\bar{x} = d_i$, $i = 1, \ldots, k$, it follows that $\sum\limits_{i=1}^{k-1} s_id_i - d_k = 0$ and
$B_kx - d_k = 0$ for any x satisfying $B_ix = d_i$, $i = 1, \ldots, k - 1$.

Hence the equality constraint $B_k x = d_k$ is redundant and can be dropped from the minimization problem without changing the solution $\bar{x}$. Then, once we have established the theorem for the linearly independent rows of B, we can reintroduce the linearly dependent row B_k (without changing the minimization problem) and set $\bar{v}_k = 0$ in the saddlepoint problem.

By 2 above, there exist $\bar{r}_0 \in R$, $\bar{r} \in R^m$, $\bar{s} \in R^k$, $(\bar{r}_0, \bar{r}) \geq 0$, $(\bar{r}_0, \bar{r}, \bar{s}) \neq 0$, which satisfy $\bar{r}g(\bar{x}) = 0$ and solve the saddlepoint problem of 2. If $\bar{r}_0 > 0$, then $\bar{u} = \bar{r}/\bar{r}_0$, $\bar{v} = \bar{s}/\bar{r}_0$ solve the saddlepoint problem of the present theorem, and we are done. Suppose $\bar{r}_0 = 0$. Then since $\bar{r}g(\bar{x}) = 0$ and $B\bar{x} - d = 0$, we have by the second inequality of the saddlepoint problem of 2 that

$$0 \leq \bar{r}g(x) + \bar{s}(Bx - d) \qquad \text{for all } x \in R^n$$

which contradicts (ii) above, if $\bar{r} \geq 0$. Now suppose that $\bar{r} = 0$, then $\bar{s} \neq 0$ and $\bar{s}(Bx - d) \geq 0$ for all x in R^n. Hence (see last part of proof of 4.2.4) $B'\bar{s} = 0$, which contradicts the assumption that the rows of B are linearly independent. Thus $\bar{r}_0 > 0$. ∎

Chapter Six

Differentiable Convex and Concave Functions

In this chapter we give some of the properties of differentiable and twice-differentiable convex and concave functions. Appendix D summarizes the results of differentiable and twice-differentiable functions which are needed in this chapter.

1. Differentiable convex and concave functions

Let θ be a numerical function defined on an open set Γ in R^n. We recall from Appendix D that if θ is differentiable at $\bar{x} \in \Gamma$, then

$$
\left.\begin{array}{c} x \in R_n \\ \\ \bar{x} + x \in \Gamma \end{array}\right\} \Rightarrow \left\{\begin{array}{l} \theta(\bar{x} + x) = \theta(\bar{x}) \\ \qquad + \nabla\theta(\bar{x})x \\ \qquad\quad + \alpha(\bar{x},x)\|x\| \\ \lim_{x \to 0} \alpha(\bar{x},x) = 0 \end{array}\right.
$$

where $\nabla\theta(\bar{x})$ is the n-dimensional gradient vector of θ at $\bar{x}$ whose n components are the partial derivatives of θ with respect to $x_1, \ldots, x_n$ evaluated at $\bar{x}$, and α is a numerical function of x.

1   **Theorem**

Let θ be a numerical function defined on an open set $\Gamma \subset R^n$ and let θ be differentiable at $\bar{x} \in \Gamma$. If θ is convex at $\bar{x} \in \Gamma$, then

$$\theta(x) - \theta(\bar{x}) \geqq \nabla\theta(\bar{x})(x - \bar{x})$$
$$\text{for each } x \in \Gamma$$

If θ is concave at $\bar{x} \in \Gamma$, then

$$\theta(x) - \theta(\bar{x}) \leqq \nabla\theta(\bar{x})(x - \bar{x})$$
$$\text{for each } x \in \Gamma$$

PROOF Let θ be convex at $\bar{x}$. Since Γ is open, there exists an open ball $B_\delta(\bar{x})$ around $\bar{x}$, which is contained in Γ. Let $x \in \Gamma$, and let $x \neq \bar{x}$. Then for some μ, such that $0 < \mu < 1$ and $\mu < \delta/\|x - \bar{x}\|$, we have that

$$\hat{x} = \bar{x} + \mu(x - \bar{x}) = (1 - \mu)\bar{x} + \mu x \in B_\delta(\bar{x}) \subset \Gamma$$

Since θ is convex at $\bar{x}$, it follows from *4.1.1*, the convexity of $B_\delta(\bar{x})$ (see *3.1.7*), and the fact that $\hat{x} \in B_\delta(\bar{x})$, that for $0 < \lambda \leq 1$

$$(1 - \lambda)\theta(\bar{x}) + \lambda\theta(\hat{x}) \geq \theta[(1 - \lambda)\bar{x} + \lambda\hat{x}]$$

or

$$\theta(\hat{x}) - \theta(\bar{x}) \geqq \frac{\theta[\bar{x} + \lambda(\hat{x} - \bar{x})] - \theta(\bar{x})}{\lambda}$$

$$= \frac{\lambda\nabla\theta(\bar{x})(\hat{x} - \bar{x}) + \alpha[\bar{x}, \lambda(\hat{x} - \bar{x})]\lambda\|\hat{x} - \bar{x}\|}{\lambda}$$

$$= \nabla\theta(\bar{x})(\hat{x} - \bar{x}) + \alpha[\bar{x}, \lambda(\hat{x} - \bar{x})]\|\hat{x} - \bar{x}\|$$

Since

$$\lim_{\lambda \to 0} \alpha[\bar{x}, \lambda(\hat{x} - \bar{x})] = 0$$

taking the limit of the previous expression as λ approaches zero gives

$$\theta(\hat{x}) - \theta(\bar{x}) \geqq \nabla\theta(\bar{x})(\hat{x} - \bar{x})$$

Since θ is convex at $\bar{x}$, since $\hat{x} \in \Gamma$, and since $\hat{x} = (1 - \mu)\bar{x} + \mu x$, we have by *4.1.1* that

$$\mu[\theta(x) - \theta(\bar{x})] \geqq \theta(\hat{x}) - \theta(\bar{x})$$

But since

$$\hat{x} - \bar{x} = \mu(x - \bar{x})$$

and $\mu > 0$, the last three relations give

$$\theta(x) - \theta(\bar{x}) \geqq \nabla\theta(\bar{x})(x - \bar{x})$$

The proof for the concave case follows in a similar way to the above by using *4.1.2* instead of *4.1.1*. ∎

2   ### Theorem

Let θ be a numerical differentiable function on an open convex set $\Gamma \subset R^n$. θ is convex on Γ if and only if

$$\theta(x^2) - \theta(x^1) \geqq \nabla\theta(x^1)(x^2 - x^1) \qquad \text{for each } x^1, x^2 \in \Gamma$$

θ is concave on Γ if and only if

$$\theta(x^2) - \theta(x^1) \leqq \nabla\theta(x^1)(x^2 - x^1) \qquad \text{for each } x^1, x^2 \in \Gamma$$

PROOF

(*Necessity*)   Since θ is convex (concave) at each $x^1 \in \Gamma$, this part of the proof follows from Theorem 1 above.

(*Sufficiency*)   We shall establish the result for the convex case.   The concave case follows in a similar way.   Let $x^1, x^2 \in \Gamma$, and let $0 \leq \lambda \leq 1$. Since Γ is convex, $(1 - \lambda)x^1 + \lambda x^2 \in \Gamma$.   We have then

$$\theta(x^1) - \theta[(1 - \lambda)x^1 + \lambda x^2] \geq \lambda \nabla \theta[(1 - \lambda)x^1 + \lambda x^2](x^1 - x^2)$$

$$\theta(x^2) - \theta[(1 - \lambda)x^1 + \lambda x^2] \geq -(1 - \lambda)\nabla \theta[(1 - \lambda)x^1 + \lambda x^2](x^1 - x^2)$$

Multiplying the first inequality by $(1 - \lambda)$, the second one by λ, and adding gives

$$(1 - \lambda)\theta(x^1) + \lambda \theta(x^2) \geq \theta[(1 - \lambda)x^1 + \lambda x^2] \quad \blacksquare$$

A geometric interpretation of the above results can be given as follows.   For a differentiable convex function θ on Γ, the linearization $\theta(\bar{x}) + \nabla \theta(\bar{x})(x - \bar{x})$ at $\bar{x}$ never overestimates $\theta(x)$ for any x in Γ, see Fig. 6.1.1.   For a differentiable concave function θ on Γ, the linearization $\theta(\bar{x}) + \nabla \theta(\bar{x})(x - \bar{x})$ at $\bar{x}$ never underestimates $\theta(x)$ for any x in Γ, see Fig. 6.1.2.

3   **Theorem**

Let θ be a numerical differentiable function on an open convex set $\Gamma \subset R^n$.   A necessary and sufficient condition that θ be convex (concave) on Γ is that for each $x^1, x^2 \in \Gamma$

$$[\nabla \theta(x^2) - \nabla \theta(x^1)](x^2 - x^1) \geq 0 \qquad (\leq 0)$$

Fig. 6.1.1 Linearization of a convex function θ never overestimates the function.

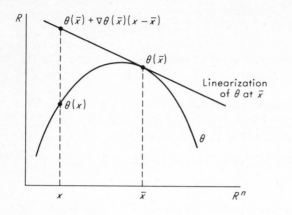

Fig. 6.1.2 Linearization of a concave function θ never underestimates the function.

PROOF We shall establish the theorem for the convex case. The concave case is similar.

(*Necessity*) Let θ be convex on Γ and let $x^1, x^2 \in \Gamma$. By Theorem 2 we have that

$$\theta(x^2) - \theta(x^1) - \nabla\theta(x^1)(x^2 - x^1) \geqq 0$$

and

$$\theta(x^1) - \theta(x^2) - \nabla\theta(x^2)(x^1 - x^2) \geqq 0$$

Adding these two inequalities gives

$$[\nabla\theta(x^2) - \nabla\theta(x^1)](x^2 - x^1) \geqq 0$$

(*Sufficiency*) Let $x^1, x^2 \in \Gamma$. Then for $0 \leqq \lambda \leqq 1$, $(1 - \lambda)x^1 + \lambda x^2 \in \Gamma$. Now by the mean-value theorem *D.2.1* we have for some $\bar{\lambda}$, $0 < \bar{\lambda} < 1$

$$\theta(x^2) - \theta(x^1) = \nabla\theta[x^1 + \bar{\lambda}(x^2 - x^1)](x^2 - x^1)$$

But by assumption

$$\{\nabla\theta[x^1 + \bar{\lambda}(x^2 - x^1)] - \nabla\theta(x^1)\}\bar{\lambda}(x^2 - x^1) \geqq 0$$

or

$$\nabla\theta[x^1 + \bar{\lambda}(x^2 - x^1)](x^2 - x^1) \geqq \nabla\theta(x^1)(x^2 - x^1)$$

Hence

$$\theta(x^2) - \theta(x^1) \geqq \nabla\theta(x^1)(x^2 - x^1)$$

and by Theorem 2 above, θ is convex on Γ. ∎

If f is an n-dimensional function on $\Gamma \subset R^n$, and $[f(x^2) - f(x^1)]$

$(x^2 - x^1) \geq 0$ for all $x^1, x^2 \in \Gamma$, then f is said to be *monotone* on Γ. It is seen from the above theorem that a differentiable numerical function on the open convex set $\Gamma \subset R^n$ is convex if and only if $\nabla\theta$ is monotone on Γ. There exists an extensive literature on monotone functions [Zarantonello 60, Browder 66, Minty 64, Opial 67], which is involved mainly with the solution of nonlinear equations in more general spaces than R^n. Some use of monotone functions in nonlinear programming has been made [Karamardian 66, Rockafellar 67b], but the full power of the theory of monotone functions has not, it seems, been exploited in nonlinear programming.

2. Differentiable strictly convex and concave functions

All the results of the previous section extend directly to strictly convex and strictly concave functions by changing the inequalities $\geq$ and $\leq$ to strict inequalities $>$ and $<$. In particular we have the following results.

1   **Theorem**

Let θ be a numerical function defined on an open set $\Gamma \subset R^n$, and let θ be differentiable at $\bar{x} \in \Gamma$. If θ is strictly convex at $\bar{x} \in \Gamma$, then

$$\theta(x) - \theta(\bar{x}) > \nabla\theta(\bar{x})(x - \bar{x}) \qquad \text{for each } x \in \Gamma, \, x \neq \bar{x}$$

If θ is strictly concave at $\bar{x} \in \Gamma$, then

$$\theta(x) - \theta(\bar{x}) < \nabla\theta(\bar{x})(x - \bar{x}) \qquad \text{for each } x \in \Gamma, \, x \neq \bar{x}$$

2   **Theorem**

Let θ be a numerical differentiable function on an open convex set $\Gamma \subset R^n$. θ is strictly convex on Γ if and only if

$$\theta(x^2) - \theta(x^1) > \nabla\theta(x^1)(x^2 - x^1) \qquad \text{for each } x^1, x^2 \in \Gamma, \, x^1 \neq x^2$$

θ is strictly concave on Γ if and only if

$$\theta(x^2) - \theta(x^1) < \nabla\theta(x^1)(x^2 - x^1) \qquad \text{for each } x^1, x^2 \in \Gamma, \, x^1 \neq x^2$$

3   **Theorem**

Let θ be a numerical differentiable function on an open convex set $\Gamma \subset R^n$. A necessary and sufficient condition that θ be strictly convex (concave) on Γ is that for each $x^1, x^2 \in \Gamma, \, x^1 \neq x^2$

$$[\nabla\theta(x^2) - \nabla\theta(x^1)](x^2 - x^1) > 0 \qquad (<0)$$

The proofs of Theorems *2* and *3* above are essentially identical to the proofs of Theorems *6.1.2* and *6.1.3* respectively and are left to the reader.   The proof of Theorem *1* is not only different from the proof of Theorem *6.1.1* but also makes use of Theorem *6.1.1*.   We give this proof below.

PROOF OF THEOREM 1   Let θ be strictly convex at $\bar{x}$.   Then

4   $\theta[(1 - \lambda)\bar{x} + \lambda x] < (1 - \lambda)\theta(\bar{x}) + \lambda\theta(x), \qquad$ for all $x \in \Gamma$, $x \neq \bar{x}$, $0 < \lambda < 1$, and $(1 - \lambda)\bar{x} + \lambda x \in \Gamma$

Since θ is convex at $\bar{x}$, it follows from Theorem *6.1.1* that

5   $\theta(x) - \theta(\bar{x}) \geqq \nabla\theta(\bar{x})(x - \bar{x}) \qquad$ for all $x \in \Gamma$

We will now show that if equality holds in *5* for some x in Γ which is distinct from $\bar{x}$, then a contradiction ensues.   Let equality hold in *5* for $x = \tilde{x}$, $\tilde{x} \in \Gamma$ and $\tilde{x} \neq \bar{x}$.   Then

6   $\theta(\tilde{x}) = \theta(\bar{x}) + \nabla\theta(\bar{x})(\tilde{x} - \bar{x})$

From *4* and *6* we have that

$\theta[(1 - \lambda)\bar{x} + \lambda\tilde{x}] < (1 - \lambda)\theta(\bar{x}) + \lambda\theta(\bar{x}) + \lambda\nabla\theta(\bar{x})(\tilde{x} - \bar{x}) \qquad$ for $0 < \lambda < 1$ and $(1 - \lambda)\bar{x} + \lambda\tilde{x} \in \Gamma$

or

7   $\theta[(1 - \lambda)\bar{x} + \lambda\tilde{x}] < \theta(\bar{x}) + \lambda\nabla\theta(\bar{x})(\tilde{x} - \bar{x}) \qquad$ for $0 < \lambda < 1$ and $(1 - \lambda)\bar{x} + \lambda\tilde{x} \in \Gamma$

But by applying Theorem *6.1.1* again to the points $x(\lambda) = (1 - \lambda)\bar{x} + \lambda\tilde{x}$, which are in Γ, and $\bar{x}$, we obtain that

$\theta[(1 - \lambda)\bar{x} + \lambda\tilde{x}] \geqq \theta(\bar{x}) + \lambda\nabla\theta(\bar{x})(\tilde{x} - \bar{x}) \qquad$ for $0 < \lambda < 1$, and $(1 - \lambda)\bar{x} + \lambda\tilde{x} \in \Gamma$

which contradicts *7*, since for small λ, $(1 - \lambda)\bar{x} + \lambda\tilde{x} \in \Gamma$ because Γ is open.   Hence equality cannot hold in *5* for any x in Γ which is distinct from $\bar{x}$, and the theorem is established for the convex case.   The concave case is similarly proved.   ∎

3.   Twice-differentiable convex and concave functions

Let θ be a numerical function defined on some open set Γ in R^n.   We recall from Appendix D that if θ is twice differentiable at $\bar{x} \in \Gamma$, then

$$\left.\begin{array}{l} x \in R^n \\ \\ \bar{x} + x \in \Gamma \end{array}\right\} \Rightarrow \left\{\begin{array}{l} \theta(\bar{x} + x) = \theta(\bar{x}) + \nabla\theta(\bar{x})x + \dfrac{x\nabla^2\theta(\bar{x})x}{2} + \beta(\bar{x},x)(\|x\|)^2 \\ \\ \lim_{x \to 0} \beta(\bar{x},x) = 0 \end{array}\right.$$

where $\nabla\theta(\bar{x})$ is the n-dimensional gradient vector of θ at $\bar{x}$, and $\nabla^2\theta(\bar{x})$ is the $n \times n$ Hessian matrix of θ at $\bar{x}$ whose ijth element is $\partial^2\theta(\bar{x})/\partial x_i \, \partial x_j$, $i, j = 1, \ldots, n$.

1   **Theorem**

Let θ be a numerical function defined on an open set $\Gamma \subset R^n$, and let θ be twice differentiable at $\bar{x} \in \Gamma$. If θ is convex at $\bar{x}$, then $\nabla^2\theta(\bar{x})$ is positive semidefinite, that is,

$$y\nabla^2\theta(\bar{x})y \geqq 0 \qquad \text{for all } y \in R^n$$

If θ is concave at $\bar{x}$, then $\nabla^2\theta(\bar{x})$ is negative semidefinite, that is,

$$y\nabla^2\theta(\bar{x})y \leqq 0 \qquad \text{for all } y \in R^n$$

PROOF   Let $y \in R^n$. Because Γ is open, there exists a $\bar{\lambda} > 0$ such that

$$\bar{x} + \lambda y \in \Gamma \qquad \text{for } 0 < \lambda < \bar{\lambda}$$

By Theorem *6.1.1* it follows that

$$\theta(\bar{x} + \lambda y) - \theta(\bar{x}) - \lambda\nabla\theta(\bar{x})y \geqq 0 \qquad \text{for } 0 < \lambda < \bar{\lambda}$$

But since θ is twice differentiable at $\bar{x}$

$$\theta(\bar{x} + \lambda y) - \theta(\bar{x}) - \lambda\nabla\theta(\bar{x})y = \frac{(\lambda)^2 y\nabla^2\theta(\bar{x})y}{2} + (\lambda)^2\beta(\bar{x},\lambda y)(\|y\|)^2$$

Hence

$$\frac{y\nabla^2\theta(\bar{x})y}{2} + \beta(\bar{x},\lambda y)(\|y\|)^2 \geqq 0 \qquad \text{for } 0 < \lambda < \bar{\lambda}$$

Taking the limit as λ approaches zero, and recalling that $\lim\limits_{\lambda \to 0} \beta(\bar{x},\lambda y) = 0$, we get that

$$y\nabla^2\theta(\bar{x})y \geqq 0$$

The concave case is established in a similar way. ■

2   **Theorem**

Let θ be a numerical twice-differentiable function on an open convex set $\Gamma \subset R^n$. θ is convex on Γ if and only if $\nabla^2\theta(x)$ is positive semidefinite on Γ, that is, for each $x \in \Gamma$

$$y\nabla^2\theta(x)y \geqq 0 \qquad \text{for all } y \in R^n$$

θ is concave on Γ if and only if $\nabla^2\theta(x)$ is negative semidefinite on Γ, that is, for each $x \in \Gamma$

$$y\nabla^2\theta(x)y \leqq 0 \qquad \text{for all } y \in R^n$$

PROOF

(Necessity)   Since θ is convex (concave) at each $x \in \Gamma$, this part of the proof follows from Theorem *1* above.

(Sufficiency)   By Taylor's theorem *D.2.2* we have that for any $x^1, x^2 \in \Gamma$

$$\theta(x^2) - \theta(x^1) - \nabla\theta(x^1)(x^2 - x^1) = \frac{(x^2 - x^1)\nabla^2\theta[x^1 + \delta(x^2 - x^1)](x^2 - x^1)}{2}$$

for some δ, $0 < \delta < 1$.   But the right-hand side of the above equality is nonnegative (nonpositive), because $\nabla^2\theta(x)$ is positive (negative) semi-definite on Γ, and $x^1 + \delta(x^2 - x^1) \in \Gamma$.   Hence the left-hand side is nonnegative (nonpositive), and by Theorem *6.1.2* θ is convex (concave) on Γ. ∎

4.   Twice-differentiable strictly convex and concave functions

Not all the results of the previous section extend to strictly convex and strictly concave functions by replacing inequalities by strict inequalities. In fact we begin by establishing the following partially negative result.

1   **Theorem**

   *Let θ be a numerical function defined on an open set $\Gamma \subset R^n$, and let θ be twice differentiable at $\bar{x} \in \Gamma$.   If θ is strictly convex at $\bar{x}$, then $\nabla^2\theta(\bar{x})$ is positive semidefinite, but **not** necessarily positive definite; that is, it is not necessarily true that*

$$y\nabla^2\theta(\bar{x})y > 0 \qquad \text{for all } y \in R^n, y \neq 0$$

*If θ is strictly concave at $\bar{x}$, then $\nabla^2\theta(\bar{x})$ is negative semidefinite, but **not** necessarily negative definite; that is, it is not necessarily true that*

$$y\nabla^2\theta(\bar{x})y < 0 \qquad \text{for all } y \in R^n, y \neq 0$$

PROOF   If θ is strictly convex at $\bar{x}$, then θ is convex at $\bar{x}$, and by Theorem *6.3.1* $\nabla^2\theta(\bar{x})$ is positive semidefinite.   That $\nabla^2\theta(\bar{x})$ is not necessarily positive definite can be seen from the counterexample $\theta(x) = (x)^4$, $x \in R$. θ is strictly convex on R but $\nabla^2\theta(x) = 12(x)^2$ is not positive definite since $\nabla^2\theta(0) = 0$.   The concave case is established similarly. ∎

Theorem

*Let θ be a numerical twice-differentiable function on an open convex set $\Gamma \subset R^n$. A **non**necessary but sufficient condition that θ be strictly convex on Γ is that $\nabla^2\theta(x)$ be positive definite on Γ; that is, for each $x \in \Gamma$*

$$y\nabla^2\theta(x)y > 0 \qquad for\ all\ y \in R^n,\ y \neq 0$$

*A **non**necessary but sufficient condition that θ be strictly concave on Γ is that $\nabla^2\theta(x)$ be negative definite on Γ; that is, for each $x \in \Gamma$*

$$y\nabla^2\theta(x)y < 0 \qquad for\ all\ y \in R^n,\ y \neq 0$$

PROOF   The **non**necessity follows from Theorem *1* above.   The sufficiency proof is essentially identical to the sufficiency proof of Theorem *6.3.2*.   ∎

Chapter Seven

Optimality Criteria in Nonlinear Programming With Differentiability

In Chap. 5 we developed optimality criteria without the use of any differentiability assumptions on the functions entering into the nonlinear programming problem. Many problems (for example linear programming and quadratic programming) involve differentiable functions. It is important then to develop optimality criteria that take advantage of this property. These criteria are merely extensions of the well-known and often abused optimality criterion of the classical calculus of "setting the derivatives equal to zero."

As we did in Chap. 5, we shall develop necessary and sufficient optimality criteria. For the sufficient optimality criteria we shall need differentiability and convexity (in *5.3.1* no convexity was needed). For the necessary optimality criteria we shall need differentiability, and, depending on which criterion we are talking about, we shall or we shall not need a constraint qualification. Note that we do not need any convexity requirements in order to establish necessary optimality criteria here. This is unlike the necessary optimality criteria of Sec. *5.4*, where convexity was crucially needed.

Again, as is the case of Chap. 5, the main sufficient optimality criteria here are straightforward to establish. Only simple inequalities relating convex functions and their gradients are needed. The necessary optimality criteria, on the other hand, need more sophisticated arguments that involve theorems of the alternative, and also, some of them require some sort of constraint qualification.

Nonlinear equality constraints will be treated cursorily in this chapter. A subsequent chapter, Chap. 11, will be devoted to programming problems with nonlinear equality constraints.

1. The minimization problems, and the Fritz John and Kuhn-Tucker stationary-point problems

The optimality criteria of this chapter relate the solutions of a minimization problem, a local minimization problem, and two stationary-point problems (the Fritz John and Kuhn-Tucker problems) to each other. The minimization and the local minimization problems considered here will be the same as the corresponding problems of Chap. 5, that is, problems *5.1.1* and *5.1.2*, with the added differentiability assumption. The Fritz John and Kuhn-Tucker problems of this chapter (*3* and *4* below) follow from the Fritz John and Kuhn-Tucker saddlepoint problems *5.1.3* and *5.1.4* if differentiability is assumed, and conversely the Fritz John and Kuhn-Tucker saddlepoint problems *5.1.3* and *5.1.4* follow from the Fritz John and Kuhn-Tucker problems (*3* and *4* below) if convexity is assumed (see *7.3.8* below).

Let X^0 be an open set in R^n, and let θ and g be respectively a numerical function and an m-dimensional vector function both defined on X^0. (In many nonlinear programming problems X^0 is R^n.)

1   **The minimization problem (MP)**

Find an $\bar{x}$, if it exists, such that

$$\theta(\bar{x}) = \min_{x \in X} \theta(x) \qquad \bar{x} \in X = \{x \mid x \in X^0, g(x) \leq 0\} \qquad \text{(MP)}$$

2   **The local minimization problem (LMP)**

Find an $\bar{x}$ in X, if it exists, such that for some open ball $B_\delta(\bar{x})$ around $\bar{x}$ with radius $\delta > 0$

$$x \in B_\delta(\bar{x}) \cap X \Rightarrow \theta(x) \geq \theta(\bar{x}) \qquad \text{(LMP)}$$

3   **The Fritz John stationary-point problem (FJP)**

Find $\bar{x} \in X^0$, $\bar{r}_0 \in R$, $\bar{r} \in R^m$ if they exist, such that

$$\left. \begin{array}{l} \nabla_x \phi(\bar{x}, \bar{r}_0, \bar{r}) = 0 \\ \nabla_r \phi(\bar{x}, \bar{r}_0, \bar{r}) \leq 0 \\ \bar{r} \nabla_r \phi(\bar{x}, \bar{r}_0, \bar{r}) = 0 \\ (\bar{r}_0, \bar{r}) \geq 0 \\ \phi(x, r_0, r) = r_0 \theta(x) + rg(x) \end{array} \right\} \quad \begin{array}{c} or \\ \\ equivalently \end{array} \quad \left\{ \begin{array}{l} \bar{r}_0 \nabla \theta(\bar{x}) + \bar{r} \nabla g(\bar{x}) = 0 \\ g(\bar{x}) \leq 0 \\ \bar{r} g(\bar{x}) = 0 \\ (\bar{r}_0, \bar{r}) \geq 0 \end{array} \right.$$

$$\text{(FJP)}$$

(It is implicit in the above statement that θ and g are differentiable at $\bar{x}$.)

4   **The Kuhn-Tucker stationary-point problem (KTP)**

Find $\bar{x} \in X^0$, $\bar{u} \in R^m$ if they exist, such that

$$
\left.\begin{array}{c}
\nabla_x \psi(\bar{x},\bar{u}) = 0 \\
\nabla_u \psi(\bar{x},\bar{u}) \leqq 0 \\
\bar{u}\nabla_u \psi(\bar{x},)\bar{u} = 0 \\
\bar{u} \geqq 0 \\
\psi(x,u) = \theta(x) + ug(x)
\end{array}\right\}
\quad \begin{array}{c} or \\ equivalently \end{array} \quad
\left\{\begin{array}{c}
\nabla\theta(\bar{x}) + \bar{u}\nabla g(\bar{x}) = 0 \\
g(\bar{x}) \leqq 0 \\
\bar{u}g(\bar{x}) = 0 \\
\bar{u} \geqq 0
\end{array}\right.
$$

(KTP)

(Again, it is implicit in the above statement that θ and g are differentiable at $\bar{x}$.)

5   **Remark**

If $(\bar{x},\bar{r}_0,\bar{r})$ is a solution of FJP *3*, and $\bar{r}_0 > 0$, then $(\bar{x},\bar{r}/\bar{r}_0)$ is a solution of KTP *4*. Conversely, if $(\bar{x},\bar{u})$ is a solution of KTP *4*, then $(\bar{x},1,\bar{u})$ is a solution of FJP *3*.

6   **Remark**

The Lagrangian functions $\phi(x,r_0,r)$ and $\psi(x,u)$ defined above are precisely the same Lagrangian functions defined in Chap. 5 (see *5.1.3* and *5.1.4*).

2.   Sufficient optimality criteria

The sufficient optimality criteria developed here (*1* and *2* below), unlike the sufficient optimality criteria *5.3.1*, depend heavily on convexity. Their derivation, however, is quite straightforward.

1   **Sufficient optimality theorem [Kuhn-Tucker 51]**

Let $\bar{x} \in X^0$, let X^0 be open, and let θ and g be *differentiable and convex* at $\bar{x}$. If $(\bar{x},\bar{u})$ is a solution of KTP *7.1.4*, then $\bar{x}$ is a solution of MP *7.1.1*. If $(\bar{x},\bar{r}_0,\bar{r})$ is a solution of FJP *7.1.3*, and $\bar{r}_0 > 0$, then $\bar{x}$ is a solution of MP *7.1.1*.

PROOF   The second statement of the theorem follows trivially from the first statement by Remark *7.1.5*.

Let $(\bar{x}, \bar{u})$ be a solution of KTP.   We have for any x in X that

$$\theta(x) - \theta(\bar{x}) \geqq \nabla\theta(\bar{x})(x - \bar{x}) \qquad \text{(by convexity and differentiability of } \theta \text{ at } \bar{x} \text{ and } 6.1.1\text{)}$$

$$= -\bar{u}\nabla g(\bar{x})(x - \bar{x}) \qquad [\text{since } \nabla\theta(\bar{x}) = -\bar{u}\nabla g(\bar{x})]$$

$$\geqq \bar{u}[g(\bar{x}) - g(x)] \qquad \text{(by convexity and differentiability of } g \text{ at } \bar{x} \text{ and } 6.1.1, \text{ and by } \bar{u} \geqq 0\text{)}$$

$$= -\bar{u}g(x) \qquad [\text{since } \bar{u}g(\bar{x}) = 0]$$

$$\geqq 0 \qquad [\text{since } \bar{u} \geqq 0 \text{ and } g(x) \leqq 0]$$

Hence

$$\theta(x) \geqq \theta(\bar{x}) \qquad \text{for any } x \in X$$

Since $g(\bar{x}) \leqq 0$, $\bar{x}$ is in X, and hence

$$\theta(\bar{x}) = \min_{x \in X} \theta(x) \qquad \text{and} \qquad \bar{x} \in X \quad \blacksquare$$

It should be remarked here that because of the convexity requirements on g, nonlinear equality constraints of the type $h(x) = 0$ cannot be handled by the above theorem by replacing them by two inequalities $h(x) \leqq 0$ and $-h(x) \leqq 0$.   However, linear equality constraints can be handled by this procedure.

2   Problem

Let $\bar{x} \in X^0$, let X^0 be open, let θ and g be differentiable and convex at $\bar{x}$, let B be a given $k \times n$ matrix, and let d be a given k-dimensional vector.   Show that if $(\bar{x}, \bar{u}, \bar{v})$, $\bar{x} \in X^0$, $\bar{u} \in R^m$, $\bar{v} \in R^k$ is a solution of the following Kuhn-Tucker problem

$$\nabla\theta(\bar{x}) + \bar{u}\nabla g(\bar{x}) + B'\bar{v} = 0$$

$$g(\bar{x}) \leqq 0$$

$$B\bar{x} = d$$

$$\bar{u}g(\bar{x}) = 0$$

$$\bar{u} \geqq 0$$

then

$$\theta(\bar{x}) = \min_{x \in X} \theta(x) \qquad \bar{x} \in X = \{x \mid x \in X^0, g(x) \leqq 0, Bx = d\}$$

An interesting case not covered by Theorem 1 above is the case when $(\bar{x}, \bar{r}_0, \bar{r})$ is a solution of FJP $7.1.3$ but the requirement that $\bar{r}_0 > 0$

is not made, in order to guarantee that $\bar{x}$ be a solution of MP *7.1.1.* This is given by the following theorem, where instead of requiring that $\bar{r}_0 > 0$, we require that g be strictly convex at $\bar{x}$.

3   **Sufficient optimality theorem**

        Let $\bar{x} \in X^0$, let X^0 be open, let θ be differentiable and convex at $\bar{x}$, and let g be differentiable and strictly convex at $\bar{x}$. If $(\bar{x}, \bar{r}_0, \bar{r})$ is a solution of FJP 7.1.3, then $\bar{x}$ solves MP 7.1.1.

PROOF   Let $(\bar{x}, \bar{r}_0, \bar{r})$ solve FJP.   Let

$$I = \{i \mid g_i(\bar{x}) = 0\} \quad J = \{i \mid g_i(\bar{x}) < 0\} \quad I \cup J = \{1, 2, \ldots, m\}\dagger$$

Since $\bar{r} \geq 0$, $g(\bar{x}) \leq 0$, and $\bar{r}g(\bar{x}) = 0$, we have that

$$\bar{r}_i g_i(\bar{x}) = 0 \quad \text{for } i = 1, \ldots, m$$

and hence

$$\bar{r}_i = 0 \quad \text{for } i \in J$$

Since $\bar{r}_0 \nabla\theta(\bar{x}) + \bar{r}\nabla g(\bar{x}) = 0$ and $(\bar{r}_0, \bar{r}) \geq 0$, we have that

$$\bar{r}_0 \nabla\theta(\bar{x}) + \sum_{i \in I} \bar{r}_i \nabla g_i(\bar{x}) = 0$$

$$(\bar{r}_0, \bar{r}_I) \geq 0$$

It follows then by Gordan's theorem *2.4.5* that

4 $\left\langle \begin{array}{c} \nabla\theta(\bar{x})z < 0 \\ \nabla g_I(\bar{x})z < 0 \end{array} \right\rangle$ has no solution $z \in R^n$

Consequently

5 $\left\langle \begin{array}{c} 0 > \theta(x) - \theta(\bar{x}) \\ 0 \geq g_I(x) - g_I(\bar{x}) \end{array} \right\rangle$ has no solution $x \in X^0$

for if it did have a solution $\hat{x} \in X^0$, then $\hat{x} \neq \bar{x}$, and

$$0 > \theta(\hat{x}) - \theta(\bar{x}) \geq \nabla\theta(\bar{x})(\hat{x} - \bar{x}) \qquad \text{(by convexity of } \theta \text{ at } \bar{x} \text{ and } 6.1.1)$$

$$0 \geq g_I(\hat{x}) - g_I(\bar{x}) > \nabla g_I(\bar{x})(\hat{x} - \bar{x}) \qquad \text{(by strict convexity of } g \text{ at } \bar{x} \text{ and } 6.2.1)$$

---

† Sometimes the constraints $g_I(x) \leq 0$ are said to be *active constraints* at $\bar{x}$, and $g_J(x) \leq 0$ are said to be *inactive constraints* at $\bar{x}$.

which contradicts 4 if we set $z = \hat{x} - \bar{x}$. Recalling that $g_I(\bar{x}) = 0$, we have from 5 that

$$\left\langle \begin{array}{c} \theta(\bar{x}) > \theta(x) \\ 0 \geqq g_I(x) \\ 0 \geqq g_J(x) \end{array} \right\rangle \quad \text{has no solution } x \in X^0$$

Since $g(\bar{x}) \leqq 0$, $\bar{x}$ is in X, and hence $\bar{x}$ solves MP $7.1.1$. ∎

6   **Remark**

In 1 and 2 above, since $\bar{u} \geqq 0$, $g(\bar{x}) \leqq 0$, and $\bar{u}g(\bar{x}) = 0$, we have that $\bar{u}_i g_i(\bar{x}) = 0$ for $i = 1, \ldots, m$, and hence $\bar{u}_i = 0$ for

$$i \in J = \{i \mid g_i(\bar{x}) < 0\}$$

Similarly in 3 we have already shown that $\bar{r}_i = 0$ for $i \in J$. It is obvious then from the proofs above that the convexity of g_I at $\bar{x}$ is all that is needed in 1 and 2 instead of the convexity of g at $\bar{x}$ (as was assumed in 1 and 2), and that the strict convexity of g_I at $\bar{x}$ is all that is needed in 3 instead of the strict convexity of g at $\bar{x}$ (as was assumed in 3). The more restrictive assumptions were made in 1, 2, and 3 to make the statements of the theorems simpler.

3. Necessary optimality criteria

In the necessary optimality conditions to be derived now, convexity does not play any crucial role. The differentiability property of the functions is used to linearize the nonlinear programming problem, and then theorems of the alternative are used to obtain the necessary optimality conditions. Once again, to derive the more important necessary optimality conditions (7 below), constraint qualifications are needed.

We begin with the following slight generalization of Abadie's linearization lemma, which establishes the nonexistence of a solution of a system of linear inequalities whenever the local minimization problem LMP $7.1.2$ has a solution.

1   **Linearization lemma [Abadie 67]**

Let $\bar{x}$ be a solution of LMP $7.1.2$, *let X^0 be open, let θ and g be differentiable at $\bar{x}$, let*

$$V = \{i \mid g_i(\bar{x}) = 0, \text{ and } g_i \text{ is concave at } \bar{x}\}$$

and let

$W = \{i \mid g_i(\bar{x}) = 0,$ *and* g_i *is not concave at* $\bar{x}\}$

Then the system

$$\left\langle \begin{array}{c} \nabla\theta(\bar{x})z < 0 \\ \nabla g_W(\bar{x})z < 0 \\ \nabla g_V(\bar{x})z \leqq 0 \end{array} \right\rangle$$

has no solution z *in* R^n.

PROOF Let

$I = V \cup W = \{i \mid g_i(\bar{x}) = 0\}$

$J = \{i \mid g_i(\bar{x}) < 0\}$

Hence

$I \cup J = V \cup W \cup J = \{1, 2, \ldots, m\}$

Let $\bar{x}$ be a solution of LMP *7.1.2* with $\delta = \bar{\delta}$. We shall show that if z satisfies $\nabla\theta(\bar{x})z < 0$, $\nabla g_W(\bar{x})z < 0$, and $\nabla g_V(\bar{x})z \leqq 0$ then a contradiction ensues. Let z satisfy these inequalities. Then, since X^0 is open, there exists a $\hat{\delta} > 0$ such that

$\bar{x} + \delta z \in X^0 \qquad$ for $0 < \delta < \hat{\delta}$

Since θ and g are differentiable at $\bar{x}$ (see *D.1.3*), we have that for $0 < \delta < \hat{\delta}$

$\theta(\bar{x} + \delta z) = \theta(\bar{x}) + \delta\nabla\theta(\bar{x})z + \alpha_0(\bar{x},\delta z)\delta\|z\|$

$g_i(\bar{x} + \delta z) = g_i(\bar{x}) + \delta\nabla g_i(\bar{x})z + \alpha_i(\bar{x},\delta z)\delta\|z\| \qquad i = 1, \ldots, m$

where

$\lim_{\delta \to 0} \alpha_i(\bar{x},\delta z) = {'}0 \qquad$ for $i = 0, 1, \ldots, m$

    (i)   If δ is small enough (say $0 < \delta < \delta_0$), then

$\nabla\theta(\bar{x})z + \alpha_0(\bar{x},\delta z)\|z\| < 0 \qquad$ [since $\nabla\theta(\bar{x})z < 0$]

and hence

$\theta(\bar{x} + \delta z) - \theta(\bar{x}) < 0 \qquad$ for $0 < \delta < \delta_0$

    (ii)   Similarly, for $i \in W$ and δ small enough (say $0 < \delta < \delta_i$), then

$\nabla g_i(\bar{x})z + \alpha_i(\bar{x},\delta z)\|z\| < 0 \qquad$ [since $\nabla g_W(\bar{x})z < 0$]

and hence

$$g_i(\bar{x} + \delta z) - g_i(\bar{x}) < 0 \qquad \text{for } 0 < \delta < \delta_i, \, i \in W$$

(iii)   For $i \in V$, we have, since g_i is concave at $\bar{x}$ and $\nabla g_V(\bar{x})z \leqq 0$, that

$$g_i(\bar{x} + \delta z) - g_i(\bar{x}) \leqq \delta \nabla g_i(\bar{x})z \leqq 0 \qquad \text{for } 0 < \delta < \hat{\delta} \text{ and } i \in V$$

(iv)   For $i \in J$, $g_i(\bar{x}) < 0$. Hence for small enough δ (say $0 < \delta < \delta_i$), we have

$$g_i(\bar{x}) + \delta \nabla g_i(\bar{x})z + \alpha_i(\bar{x}, \delta z)\delta\|z\| < 0$$

and hence

$$g_i(\bar{x} + \delta z) < 0 \qquad \text{for } 0 < \delta < \delta_i, \, i \in J$$

Let us call $\bar{\delta}$ the minimum of all the positive numbers $\tilde{\delta}$, $\hat{\delta}$, δ_0, δ_i $(i = 1, \ldots, m)$ defined above.   Then for any δ in the interval $0 < \delta < \bar{\delta}$ we have that

$$\bar{x} + \delta z \in X^0$$

$$\bar{x} + \delta z \in B_{\tilde{\delta}}(\bar{x})$$

$$\theta(\bar{x} + \delta z) < \theta(\bar{x})$$

$$g_i(\bar{x} + \delta z) \leqq g_i(\bar{x}) = 0 \qquad \text{for } i \in I$$

$$g_i(\bar{x} + \delta z) < 0 \qquad \text{for } i \in J$$

Hence, for $0 < \delta < \bar{\delta}$, we have that $\bar{x} + \delta z \in B_{\tilde{\delta}}(\bar{x}) \cap X$ and $\theta(\bar{x} + \delta z) < \theta(\bar{x})$, which contradicts the assumption that $\bar{x}$ is a solution of LMP 7.1.2 with $\delta = \tilde{\delta}$.   Hence there exists no z in R^n satisfying $\nabla\theta(\bar{x})z < 0$, $\nabla g_W(\bar{x}) < 0$, and $\nabla g_V(\bar{x})z \leqq 0$. ∎

We are now ready to derive a series of necessary optimality criteria based on the above linearization lemma.   We begin by a generalization of a result of Abadie which, for the case of a finite number of constraints, includes the classical result of Fritz John.

2   Fritz John stationary-point necessary optimality theorem [John 48, Abadie 67]

Let $\bar{x}$ be a solution of LMP 7.1.2 or of MP 7.1.1, let X^0 be open and let θ and g be differentiable at $\bar{x}$.   Then there exists an $\bar{r}_0 \in R$ and an $\bar{r} \in R^m$ such that $(\bar{x}, \bar{r}_0, \bar{r})$ solves FJP 7.1.3, and

$$(\bar{r}_0, \bar{r}_W) \geq 0$$

where

$W = \{i|\ g_i(\bar{x}) = 0,\ and\ g_i\ is\ not\ concave\ at\ \bar{x}\}$

PROOF   If $\bar{x}$ solves MP *7.1.1*, then $\bar{x}$ solves LMP *7.1.2*.   Let

$V = \{i\ |\ g_i(\bar{x}) = 0,\ and\ g_i\ is\ concave\ at\ \bar{x}\}$

and

$J = \{i\ |\ g_i(\bar{x}) < 0\}$

By Lemma *1* above we have that

$$\left.\begin{array}{c} \nabla\theta(\bar{x})z < 0 \\ \nabla g_W(\bar{x})z < 0 \\ \nabla g_V(\bar{x})z \leqq 0 \end{array}\right\rangle \quad \text{has no solution } z \in R^n$$

Hence by Motzkin's theorem *2.4.2* there exist $\bar{r}_0, \bar{r}_W, \bar{r}_V$ such that

$$\bar{r}_0\nabla\theta(\bar{x}) + \bar{r}_W\nabla g_W(\bar{x}) + \bar{r}_V\nabla g_V(\bar{x}) = 0$$

$$(\bar{r}_0,\bar{r}_W) \geq 0 \qquad \bar{r}_V \geqq 0$$

Since $g_W(\bar{x}) = 0$ and $g_V(\bar{x}) = 0$, it follows that if we define

$$\bar{r}_J = 0 \qquad \text{and} \qquad \bar{r} = (\bar{r}_W,\bar{r}_V,\bar{r}_J)$$

then

$$\bar{r}g(\bar{x}) = \bar{r}_W g_W(\bar{x}) + \bar{r}_V g_V(\bar{x}) + \bar{r}_J g_J(\bar{x}) \doteq 0$$

$$\bar{r}_0\nabla\theta(\bar{x}) + \bar{r}\nabla g(\bar{x}) = 0$$

and

$$(\bar{r}_0,\bar{r}) \geq 0$$

Since $\bar{x}$ is in X, $g(\bar{x}) \leqq 0$.   Hence $(\bar{x},\bar{r}_0,\bar{r})$ solves FJP *7.1.3*, and $(\bar{r}_0,\bar{r}_W) \geq 0$.   ∎

REMARK   The above theorem becomes that of Abadie [Abadie 67] if we replace the concavity of g_V at $\bar{x}$ and the nonconcavity of g_W at $\bar{x}$ by the requirements that g_V be linear on R^n and that g_W be nonlinear on R^n. It might not be immediately obvious why the above result is more general than Abadie's.   One way to see this is the following: Besides $(\bar{r}_0, \bar{r}) \geq 0$, we try to establish in addition the semipositivity of the smallest subset of the components of $(\bar{r}_0, \bar{r})$.   For example if we can establish the semipositivity of $\bar{r}_0$ (and hence its positivity), then this is the sharpest necessary optimality condition we can get (see γ below). In the above theorem we establish the semipositivity of $(\bar{r}_0, \bar{r}_W)$, where W is the set of indices of the active constraints which are not concave at $\bar{x}$. Abadie establishes the semipositivity of a larger subset of components of $(\bar{r}_0, \bar{r})$, namely that of $(\bar{r}_0, \bar{r}_N)$, where N is the set of indices of the active constraints which are nonlinear on R^n. On the other hand, the above theorem becomes that of Fritz John [John 48] if we drop the result that $(\bar{r}_0, \bar{r}_W) \geqq 0$.

REMARK If X^0 is convex and g_V is concave on X^0, then the above theorem certainly holds. However, the concavity of g_V does not (unless g_V is linear) make the set $\{x \mid x \in X^0, \ g_V(x) \leqq 0\}$ convex (see *4.1.10* and *4.1.11*).

Again, just as in the case for the saddlepoint necessary optimality criteria (see Section *5.4*), there is no guarantee that $\bar{r}_0 > 0$ in Theorem *2* above. This can occur, for example, when the solution $\bar{x}$ of LMP *7.1.2* or of MP *7.1.1* is at a cusp, Fig. *7.3.1a*, or when the feasible region x is degenerate and is made up of one point only, Fig. *7.3.1b*. In cases when $\bar{r}_0 = 0$, the function θ disappears from the Fritz John problem *7.1.3*, and we have a degenerate case. As was done in Sec. *5.4*, it is possible to exclude such cases by introducing restrictions on the constraints $g(x) \leqq 0$. These restrictions are again called, as in Sec. *5.4*, *constraint qualifications*. In addition to Slater's constraint qualification *5.4.3*, Karlin's constraint qualification *5.4.4*, and the strict constraint qualification *5.4.5*, we shall introduce the Kuhn-Tucker constraint qualification *3*, the Arrow-Hurwicz-

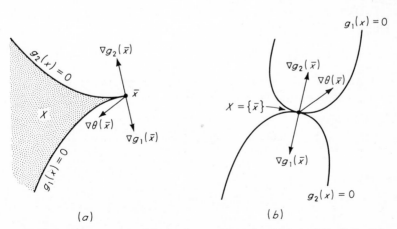

Fig. 7.3.1 Examples of minimization problems in which $\bar{r}_0$ of the Fritz John problem *7.1.3* is zero.

Uzawa constraint qualification *4*, and the reverse convex constraint qualification *5*.

3   **The Kuhn-Tucker constraint qualification [Kuhn-Tucker 51]**

Let X^0 be an open set in R^n, let g be an m-dimensional vector function defined on X^0, and let

$$X = \{x \mid x \in X^0,\, g(x) \leq 0\}$$

g is said to satisfy the *Kuhn-Tucker constraint qualification at* $\bar{x} \in X$ if g is differentiable at $\bar{x}$ and if

$$
\left.
\begin{array}{c}
y \in R^n \\
\nabla g_I(\bar{x})y \leq 0
\end{array}
\right\}
\Rightarrow
\left\{
\begin{array}{l}
\text{There exists an } n\text{-dimensional vector function} \\
e \text{ defined on the interval } [0,1] \text{ such that} \\[4pt]
a.\ \ e(0) = \bar{x} \\
b.\ \ e(\tau) \in X \text{ for } 0 \leq \tau \leq 1 \\
c.\ \ e \text{ is differentiable at } \tau = 0 \text{ and} \\[4pt]
\qquad \dfrac{de(0)}{d\tau} = \lambda y \text{ for some } \lambda > 0
\end{array}
\right.
$$

where

$$I = \{i \mid g_i(\bar{x}) = 0\}$$

4   **The Arrow-Hurwicz-Uzawa constraint qualification [Arrow et al. 61]**

Let X^0 be an open set in R^n, let g be an m-dimensional vector function defined on X^0, and let

$$X = \{x \mid x \in X^0,\, g(x) \leq 0\}$$

g is said to satisfy the *Arrow-Hurwicz-Uzawa constraint qualification at* $\bar{x} \in X$ if g is differentiable at $\bar{x}$ and if

$$
\left\langle
\begin{array}{c}
\nabla g_W(\bar{x})z > 0 \\
\nabla g_V(\bar{x})z \geq 0
\end{array}
\right\rangle
\quad \text{has a solution } z \in R^n
$$

where

$$V = \{i \mid g_i(\bar{x}) = 0, \text{ and } g_i \text{ is concave at } \bar{x}\}$$

and

$$W = \{i \mid g_i(\bar{x}) = 0, \text{ and } g_i \text{ is not concave at } \bar{x}\}$$

5   The reverse convex constraint qualification [Arrow et al. 61]

Let X^0 be an open set in R^n, let g be an m-dimensional vector function defined on X^0, and let

$$X = \{x \mid x \in X^0, g(x) \leq 0\}$$

g is said to satisfy the *reverse convex constraint qualification at* $\bar{x} \in X$ if g is differentiable at $\bar{x}$, and if for each $i \in I$ either g_i is concave at $\bar{x}$ or g_i is linear on R^n, where

$$I = \{i \mid g_i(\bar{x}) = 0\}$$

(The reason for the adjective reverse convex is that if X^0 is a convex set and g_I is concave on X^0, then the set $\{x \mid x \in X^0, g_I(x) \leq 0\}$ is not convex, but its complement relative to X^0, that is, $\{x \mid x \in X^0, g_I(x) > 0\}$, is a convex set.   For example if $X^0 = R^n$ and $g(x) = -xx + 1$, then g satisfies the reverse constraint qualification at each $\bar{x} \in \{x \mid x \in R^n, xx \geq 1\}$.)

Before deriving the fundamental Kuhn-Tucker necessary optimality conditions (7 below), we relate the three constraint qualifications just introduced to each other and to the constraint qualifications of Chap. 5.

6   **Lemma**

Let X^0 be an open set in R^n, let g be an m-dimensional vector function defined X^0, and let

$$X = \{x \mid x \in X^0, g(x) \leq 0\}$$

(i)   *If g satisfies the reverse convex constraint qualification 5 at $\bar{x}$, then g satisfies the Arrow-Hurwicz-Uzawa constraint qualification 4 at $\bar{x}$.*

(ii)   *If g satisfies the reverse convex constraint qualification 5 at $\bar{x}$, then g satisfies the Kuhn-Tucker constraint qualification 3 at $\bar{x}$.*

(iii)   *Let X^0 be convex, let g be convex on X^0, and let g be differentiable at $\bar{x}$. If g satisfies Slater's constraint qualification 5.4.3 on X^0, Karlin's constraint qualification 5.4.4 on X^0, or the strict constraint qualification 5.4.5 on X^0, then g satisfies the Arrow-Hurwicz-Uzawa constraint qualification 4 at $\bar{x}$.*

PROOF   (i)   Let g satisfy the reverse constraint qualification 5 at $\bar{x}$. Then the set W defined in 4 is empty, and $z = 0$ satisfies

$$\nabla g_V(\bar{x})z = \nabla g(\bar{x})z \geq 0$$

Hence g satisfies the Arrow-Hurwicz-Uzawa constraint qualification 4 at $\bar{x}$.

(ii)   Let g satisfy the reverse constraint qualification 5 at $\bar{x}$. Define

$$I = \{i \mid g_i(\bar{x}) = 0\} \qquad \text{and} \qquad J = \{i \mid g_i(\bar{x}) < 0\}$$

Let y be any vector in R^n satisfying

$$\nabla g_I(\bar{x})y \leq 0$$

Define

$$e(\tau) = \bar{x} + \lambda\tau y \qquad \text{for some } \lambda > 0 \text{ (to be specified below)}$$

Obviously conditions (a) and (c) of the Kuhn-Tucker constraint qualification 3 are satisfied. We will now show that condition (b) is also satisfied. Since X^0 is open, and since g_I is concave and differentiable at $\bar{x}$, we have that

$$\bar{x} + \lambda\tau y \in X^0$$

and

$$g_I[e(\tau)] = g_I(\bar{x} + \lambda\tau y) \leq g_I(\bar{x}) + \lambda\tau\nabla g_I(\bar{x})y = \lambda\tau\nabla g_I(\bar{x})y \leq 0$$

for $0 \leq \tau \leq 1$ and $0 \leq \lambda < \bar{\lambda}$. Hence for $0 \leq \lambda < \bar{\lambda}$ and $0 \leq \tau \leq 1$, we have that

$$g_I[e(\tau)] \leq 0$$

and for $i \in J$, we have that

$$g_i(\bar{x} + \lambda\tau y) = g_i(\bar{x}) + \lambda\tau[\nabla g_i(\bar{x})y + \alpha_i(\bar{x},\lambda\tau y)\|y\|] \qquad \lim_{\lambda \to 0} \alpha_i(\bar{x},\lambda\tau y) = 0$$

$$\leq g_i(\bar{x}) + \lambda\tau[\|\nabla g_i(\bar{x})\| + \alpha_i(\bar{x},\lambda\tau y)]\|y\| \qquad \text{(by } 1.3.8)$$

$$< 0 \qquad \text{for some } \lambda \quad 0 < \lambda < \bar{\lambda}$$

where the last inequality holds because $g_J(\bar{x}) < 0$ and $\lim_{\lambda \to 0} \alpha_i(\bar{x},\lambda\tau y) = 0$. Hence $g_J[e(\tau)] < 0$, for $0 \leq \tau \leq 1$. Since $g_I[e(\tau)] \leq 0$ for $0 \leq \tau \leq 1$, we have that

$$e(\tau) \in X \qquad \text{for } 0 \leq \tau \leq 1$$

and condition (b) of the Kuhn-Tucker constraint qualification 3 is satisfied.

(iii)   By Lemma $5.4.6$ we have that Slater's constraint qualification $5.4.3$ and Karlin's constraint qualification $5.4.4$ are equivalent and that the strict constraint qualification $5.4.5$ implies both Slater's and Karlin's constraint qualification. Hence we need only establish the present lemma under Slater's constraint qualification. If g satisfies Slater's constraint qualification on X^0, there exists an $\hat{x} \in X^0$ such that $g(\hat{x}) < 0$.   Since g is differentiable at $\bar{x}$, we have by 6.1.1 that

$$0 > g_I(\hat{x}) = g_I(\hat{x}) - g_I(\bar{x}) \geqq \nabla g_I(\bar{x})(\hat{x} - \bar{x})$$

where

$$I = \{i \mid g_i(\bar{x}) = 0\}$$

Hence by taking $z = \bar{x} - \hat{x}$, we have that $\nabla g_I(\bar{x})z > 0$, and the Arrow-Hurwicz-Uzawa constraint qualification 4 is satisfied at $\bar{x}$.   ∎

The results of the above lemma are summarized in Fig. 7.3.2.

We are now ready to derive the fundamental necessary optimality criterion of nonlinear programming, the Kuhn-Tucker necessary optimality criterion.   We shall establish the result under all the constraint qualifications introduced.   In view of Lemma 6 above, we need only establish the result under the Kuhn-Tucker constraint qualification 3 and the Arrow-Hurwicz-Uzawa constraint qualification 4.   (A somewhat involved proof [Abadie 67] shows that a special case of the Arrow-Hurwicz-Uzawa constraint qualification, where g_V are the linear active constraints and g_W are the nonlinear active constraints, implies the Kuhn-Tucker constraint qualification.   We take here a different and somewhat simpler approach and show that either the Arrow-Hurwicz-Uzawa or the Kuhn-Tucker constraint qualification is adequate for establishing the necessary optimality conditions we are after.)

7   **Kuhn-Tucker stationary-point necessary optimality theorem [Kuhn-Tucker 51]**

Let X^0 be an open subset of R^n, let θ and g be defined on X^0, let $\bar{x}$ solve LMP 7.1.2 or MP 7.1.1, let θ and g be differentiable at $\bar{x}$, and let g

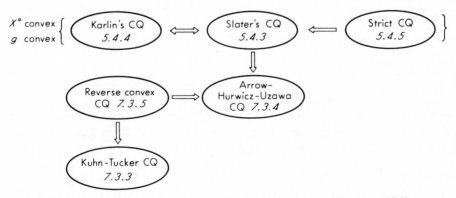

Fig. 7.3.2 Relations between the various constraint qualifications (CQ).

satisfy

 (i)   *the Kuhn-Tucker constraint qualification 3 at $\bar{x}$, or*
 (ii) *the Arrow-Hurwicz-Uzawa constraint qualification 4 at $\bar{x}$, or*
 (iii) *the reverse convex constraint qualification 5 at $\bar{x}$, or*
 (iv) *Slater's constraint qualification 5.4.3 on X^0, or*
 (v)   *Karlin's constraint qualification 5.4.4 on X^0, or*
 (vi) *the strict constraint qualification 5.4.5 on X^0.*

Then there exists a $\bar{u} \in R^m$ such that $(\bar{x}, \bar{u})$ solves KTP 7.1.4.

PROOF   In view of Lemma *6* above we need only establish the theorem under assumptions (i) or (ii).

   (i)   Let $\bar{x}$ solve LMP *7.1.2* with $\delta = \hat{\delta}$.   Let

$$I = \{i \mid g_i(\bar{x}) = 0\} \qquad J = \{i \mid g_i(\bar{x}) < 0\}$$

We have to consider two cases: the case when I is empty and the case when I is nonempty.

   $(I = \emptyset)$ Let y be any vector in R^n such that $yy = 1$.   Then

$$g_i(\bar{x} + \delta y) = g_i(\bar{x}) + \delta[\nabla g_i(\bar{x})y + \alpha_i(\bar{x}, \delta y)] \qquad \text{for } i = 1, \ldots, m$$

Since $g_i(\bar{x}) < 0$ and $\lim_{\delta \to 0} \alpha_i(\bar{x}, \delta y) = 0$, it follows that for small enough δ, say $0 < \delta < \hat{\delta} < \bar{\delta}$, $g_i(\bar{x} + \delta y) < 0$ and $\bar{x} + \delta y \in X^0$.   But since $\bar{x}$ solves LMP *7.1.2*, we have that

$$0 \leqq \theta(\bar{x} + \delta y) - \theta(\bar{x}) = \delta[\nabla \theta(\bar{x})y + \alpha(\bar{x}, \delta y)]$$

for $0 < \delta < \hat{\delta}$.   Hence

$$\nabla \theta(\bar{x})y + \alpha(\bar{x}, \delta y) \geqq 0$$

Since $\lim_{\delta \to 0} \alpha(\bar{x}, \delta y) = 0$, we have on taking the limit of the above expression as δ approaches zero that

$$\nabla \theta(\bar{x})y \geqq 0$$

Since y is an arbitrary vector in R^n satisfying $yy = 1$, we conclude from this last inequality, by taking $y = \pm e^i$, where $e^i \in R^n$ is a vector with one in the ith position and zero elsewhere, that

$$\nabla \theta(\bar{x}) = 0$$

Hence $\bar{x}$ and $\bar{u} = 0$ satisfy KTP *7.1.4*.

$(I \neq \emptyset)$ Let g satisfy the Kuhn-Tucker constraint qualification 3 at $\bar{x}$, and let $y \in R^n$ satisfy $\nabla g_I(\bar{x})y \leq 0$. Hence by 3, there exists an n-dimensional vector function e defined on $[0,1]$ such that $e(0) = \bar{x}$, $e(\tau) \in X$ for $0 \leq \tau \leq 1$, e is differentiable at $\tau = 0$, and $de(0)/d\tau = \lambda y$ for some $\lambda > 0$. Hence for $0 \leq \tau \leq 1$

$$e_i(\tau) = e_i(0) + \tau \left[\frac{de_i(0)}{d\tau} + \gamma_i(0,\tau) \right] \qquad \text{for } i = 1, \ldots, n$$

where $\lim_{\tau \to 0} \gamma_i(0,\tau) = 0$. Hence by taking τ small enough, say $0 < \tau < \hat{\tau} < 1$, we have that $e(\tau) \in B_{\bar{s}}(\bar{x})$. Since $e(\tau) \in X$ for $0 \leq \tau \leq 1$ and $\bar{x}$ solves LMP $7.1.2$, we have that

$$\theta[e(\tau)] \geq \theta[e(0)] \qquad \text{for } 0 < \tau < \hat{\tau}$$

Hence by the chain rule $D.1.6$ and the differentiability of θ at $\bar{x}$ and of e at 0, we have for $0 < \tau < \hat{\tau}$ that

$$0 \leq \theta[e(\tau)] - \theta[e(0)] = \nabla\theta[e(0)] \frac{de(0)}{d\tau} \tau + \beta(0,\tau)\tau$$

where $\lim_{\tau \to 0} \beta(0,\tau) = 0$. Hence

$$\nabla\theta[e(0)] \frac{de(0)}{d\tau} + \beta(0,\tau) \geq 0 \qquad \text{for } 0 < \tau < \hat{\tau}$$

Taking the limit as τ approaches zero gives

$$\nabla\theta[e(0)] \frac{de(0)}{d\tau} \geq 0$$

Since $e(0) = \bar{x}$ and $de(0)/d\tau = \lambda y$ for some $\lambda > 0$, we have that

$$\nabla\theta(\bar{x})y \geq 0$$

Hence we have shown that

$$\nabla g_I(\bar{x})y \leq 0 \Rightarrow \nabla\theta(\bar{x})y \geq 0$$

or that

$$\left\langle \begin{array}{c} \nabla\theta(\bar{x})y < 0 \\ \nabla g_I(\bar{x})y \leq 0 \end{array} \right\rangle \qquad \text{has no solution } y \in R^n$$

Hence by Motzkin's theorem $2.4.2$ there exists an $\bar{r}_0$ and an $\bar{r}_I$ such that

$$\bar{r}_0 \nabla\theta(\bar{x}) + \bar{r}_I \nabla g_I(\bar{x}) = 0$$

$$\bar{r}_0 \geq 0 \qquad \bar{r}_I \geq 0$$

Since $\bar{r}_0$ is a real number, $\bar{r}_0 \geq 0$ means $\bar{r}_0 > 0$.   By defining

$$\bar{u}_I = \frac{\bar{r}_I}{\bar{r}_0} \qquad \bar{u}_J = 0 \qquad \bar{u} = (\bar{u}_I, \bar{u}_J)$$

we have that

$$\nabla\theta(\bar{x}) + \bar{u}\nabla g(\bar{x}) = 0$$

$$\bar{u}g(\bar{x}) = 0$$

$$\bar{u} \geq 0$$

and since $\bar{x} \in X$, we have that

$$g(\bar{x}) \leq 0$$

Hence $(\bar{x}, \bar{u})$ solves KTP 7.1.4.

(ii)   Let $\bar{x}$ solve MP 7.1.1 or LMP 7.1.2.   Then by the Fritz John theorem 2 there exists an $\bar{r}_0 \in R$ and an $\bar{r} \in R^m$ such that $(\bar{x}, \bar{r}_0, \bar{r})$ solves FJP 7.1.3, and

$$(\bar{r}_0, \bar{r}_W) \geq 0$$

where

$$W = \{i \mid g_i(\bar{x}) = 0, \text{ and } g_i \text{ is not concave at } \bar{x}\}$$

Define

$$V = \{i \mid g_i(\bar{x}) = 0, \text{ and } g_i \text{ is concave at } \bar{x}\}$$

and

$$J = \{i \mid g_i(\bar{x}) < 0\}$$

By Remark 7.1.5 we only have to show that $\bar{r}_0 > 0$.   Since $(\bar{r}_0, \bar{r}_W) \geq 0$, we have that $\bar{r}_0 > 0$ if W is empty.   Assume now that W is nonempty. We will now show by contradiction that $\bar{r}_0 > 0$.   Suppose that $\bar{r}_0 = 0$; then since $\bar{r}_J = 0$, we have that

$$\bar{r}_W\nabla g_W(\bar{x}) + \bar{r}_V\nabla g_V(\bar{x}) = 0$$

$$\bar{r}_W \geq 0 \qquad \bar{r}_V \geq 0$$

Since g satisfies the Arrow-Hurwicz-Uzawa constraint qualification 4 at $\bar{x}$, there exists a $\bar{z} \in R^n$ such that

$$\nabla g_W(\bar{x})\bar{z} > 0$$

$$\nabla g_V(\bar{x})\bar{z} \geq 0$$

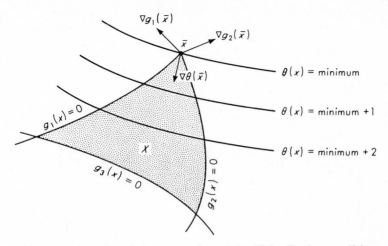

Fig. 7.3.3 A geometric interpretation of the Kuhn-Tucker conditions
7.1.4.

Premultiplying these inequalities by $\bar{r}_W$ and $\bar{r}_V$ respectively gives

$$\bar{r}_W \nabla g_W(\bar{x}) \bar{z} + \bar{r}_V \nabla g_V(\bar{x}) \bar{z} > 0$$

which contradicts the fact that

$$\bar{r}_W \nabla g_W(\bar{x}) + \bar{r}_V \nabla g_V(\bar{x}) = 0$$

Hence $\bar{r}_0 > 0$. ∎

Fig. 7.3.4 Relationship between the solutions of the local minimization problem (LMP) 7.1.2, the minimization problem (MP) 7.1.1, the Fritz John problem (FJP) 7.1.3, and the Kuhn-Tucker problem (KTSP) 7.1.4.

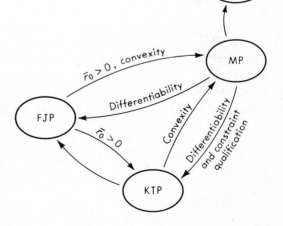

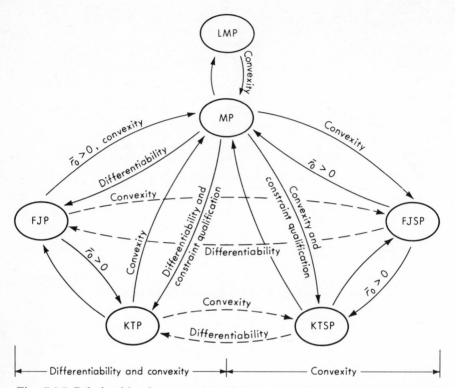

Fig. 7.3.5 Relationships between the solutions of the local minimization problem (LMP) *7.1.2*, the minimization problem (MP) *7.1.1*, the Fritz John problem (FJP) *7.1.3*, the Kuhn-Tucker problem (KTP) *7.1.4*, the Fritz John saddlepoint problem (FJSP) *5.1.3*, and the Kuhn-Tucker saddlepoint problem (KTSP) *5.1.4*.

A geometric interpretation of the Kuhn-Tucker conditions *7.1.4* can be given as follows. At $\bar{x}$, there exists a nonnegative linear combination of the gradient of the objective function $\nabla\theta(\bar{x})$ (with positive weight) and the gradients of the active constraints $\nabla g_i(\bar{x})$, $i \in I$, which is equal to zero. Figure *7.3.3* depicts such a situation. If, however, none of the constraint qualifications are satisfied, there may not exist such a nonnegative linear combination of $\nabla\theta(\bar{x})$ and $\nabla g_I(\bar{x})$. In such cases $\nabla\theta(\bar{x})$ may have a zero weight, as for example in Fig. *7.3.1*.

The various optimality criteria of this chapter are related to each other in Fig. *7.3.4*, and to the optimality criteria of Chap. 5 in Fig. *7.3.5*.

8   **Problem**

Let X^0 be an open set in R^n, and let θ and g be defined on X^0. Find the conditions under which

(i) A solution $(\bar{x}, \bar{r}_0, \bar{r})$ of the Fritz John saddlepoint problem FJSP *5.1.3* is a solution of the Fritz John problem FJP *7.1.3*, and conversely.

(ii) A solution $(\bar{x}, \bar{u})$ of the Kuhn-Tucker saddlepoint problem KTSP *5.1.4* is a solution of the Kuhn-Tucker problem KTP *7.1.4*, and conversely.

(The above relationships are indicated by dotted lines in Fig. *7.3.5*.)

9   **Problem**

Let X^0 be an open set in R^n, let θ be defined on X^0, let $\bar{x} \in X^0$, and let θ be differentiable at $\bar{x}$. Show that if

$$\theta(\bar{x}) = \min_{x \in X} \theta(x) \qquad \bar{x} \in X = \{x \mid x \in X^0, x \geqq 0\}$$

then

$$\nabla\theta(\bar{x}) \geqq 0$$

$$\bar{x} \geqq 0$$

$$\bar{x}\nabla\theta(\bar{x}) = 0$$

Under what conditions is the converse true? Give a geometric interpretation of the above conditions for the case when $X^0 = R$.

10   **Problem**

Let X^0 be an open set in R^n, let θ be defined on X^0, let $\bar{x} \in X^0$, and let θ be differentiable at $\bar{x}$. Show that if

$$\theta(\bar{x}) = \max_{x \in X} \theta(x) \qquad \bar{x} \in X = \{x \mid x \in X^0, x \geqq 0\}$$

then

$$\nabla\theta(\bar{x}) \leqq 0$$

$$\bar{x} \geqq 0$$

$$\bar{x}\nabla\theta(\bar{x}) = 0$$

Under what conditions is the converse true? Give a geometric interpretation of the above conditions for the case when $X^0 = R$.

11   **Problem (method of feasible directions [Zoutendijk 60])**

Let the assumptions of Theorem *7* hold, and let θ and g be differentiable on X^0. Suppose that we have a point $\hat{x} \in X$ at which one of the constraint qualifications of *7* holds but at which the Kuhn-Tucker conditions *7.1.4* are not satisfied. Show that there exists a *feasible direction* z in R^n such that $\hat{x} + \delta z \in X$ and $\theta(\hat{x} + \delta z) < \theta(\hat{x})$ for some

$\delta > 0$. (Hint: Use the Farkas' theorem *2.4.6*, the fact that

$$\nabla \theta(\hat{x}) + u_I \nabla g_I(\hat{x}) = 0 \qquad u_I \geqq 0$$

has no solution u_I, and the differentiability property of θ and g.) State a linear programming problem the solution of which will yield z.

12   **Problem (general Kuhn-Tucker stationary-point necessary optimality criterion)**

Let X^0 be an open set in $R^{n_1} \times R^{n_2}$, let θ, g, and h be respectively a numerical function, an m-dimensional vector function, and a k-dimensional vector function defined on X^0, let $(\bar{x}, \bar{y}) \in X^0$ solve the following general minimization problem

$$\theta(\bar{x}, \bar{y}) = \min_{(x,y) \in X} \theta(x,y) \qquad (\bar{x}, \bar{y}) \in X = \left\{ (x,y) \,\middle|\, \begin{matrix} (x,y) \in X^0, \, g(x,y) \leqq 0 \\ h(x,y) = 0, \, y \geqq 0 \end{matrix} \right\}$$

let θ, g, and h be differentiable at $(\bar{x}, \bar{y})$, and let the composite $m + 2k + n_2$-dimensional vector function f defined on X^0 by $f(x,y) = [g(x,y), h(x,y), -h(x,y), -y]$ satisfy one of the constraint qualifications *3*, *4*, or *5* at $(\bar{x}, \bar{y})$, or one of the constraint qualifications *5.4.3*, *5.4.4*, or *5.4.5* on X^0. Show that there exist $\bar{u} \in R^m$, $\bar{v} \in R^k$ such that $(\bar{x}, \bar{u}, \bar{v})$ satisfies the following general Kuhn-Tucker necessary optimality conditions

$$\nabla_x \theta(\bar{x}, \bar{y}) + \bar{u} \nabla_x g(\bar{x}, \bar{y}) + \bar{v} \nabla_x h(\bar{x}, \bar{y}) = 0$$

$$\nabla_y \theta(\bar{x}, \bar{y}) + \bar{u} \nabla_y g(\bar{x}, \bar{y}) + \bar{v} \nabla_y h(\bar{x}, \bar{y}) \geqq 0$$

$$\bar{y}[\nabla_y \theta(\bar{x}, \bar{y}) + \bar{u} \nabla_y g(\bar{x}, \bar{y}) + \bar{v} \nabla_y h(\bar{x}, \bar{y})] = 0$$

$$\bar{y} \geqq 0$$

$$g(\bar{x}, \bar{y}) \leqq 0$$

$$h(\bar{x}, \bar{y}) = 0$$

$$\bar{u} g(\bar{x}, \bar{y}) = 0$$

$$\bar{u} \geqq 0$$

(Hint: Redefine X as

$$X = \{(x,y) \mid (x,y) \in X^0, \, g(x,y) \leqq 0, \, h(x,y) \leqq 0, \, -h(x,y) \leqq 0, \, -y \leqq 0\}$$

and then use Theorem *7*.)

Chapter Eight

Duality in Nonlinear Programming

Duality plays a crucial role in the theory and computational algorithms of linear programming [Dantzig 63, Simmonard 66]. The inception of duality theory in linear programming may be traced to the classical minmax theorem of von Neumann [von Neumann 28] and was first explicitly given by Gale, Kuhn, and Tucker [Gale et al. 51]. Duality in nonlinear programming is of a somewhat later vintage, beginning with the duality results of quadratic programming [Dennis 59]. In spirit, however, duality theory in nonlinear programming is related to the reciprocal principles of the calculus of variations, which have been known since as far back as 1927 [Trefftz 27, 28, Friedrichs 29, Courant-Hilbert 53, Lax 55, Mond-Hanson 67].

In recent years there has been an extensive interest in the duality theory of nonlinear programming as evidenced by the publication of some 50 or more papers on the subject. To cover the subject of all these papers would take a book by itself. Instead we shall concentrate on a few of the more basic results which are simpler to prove and which subsume in an obvious way the duality results of linear programming.

The plan of this chapter is as follows. We first introduce the minimization problem and its dual and develop the duality results of nonlinear programming. We then apply these results to quadratic and linear programming problems.

1.   Duality in nonlinear programming

Let X^0 be an open set in R^n, and let θ and g be respectively a numerical function and an m-dimensional vector function, both defined on X^0. We define now the same minimization problem that we have been dealing with in the previous chapters and its dual.

1   **The (primal) minimization problem (MP)**

Find an $\bar{x}$, if it exists, such that

$$\theta(\bar{x}) = \min_{x \in X} \theta(x) \qquad \bar{x} \in X = \{x \in X^0, g(x) \leq 0\} \tag{MP}$$

2   **The dual (maximization) problem (DP)**

Let θ and g be differentiable on X^0. Find an $\hat{x}$ and a $\hat{u} \in R^m$, if they exist, such that

$$\left| \begin{array}{l} \psi(\hat{x},\hat{u}) = \max_{(x,u) \in Y} \psi(x,u) \\[2mm] (\hat{x},\hat{u}) \in Y = \{(x,u) \mid x \in X^0,\ u \in R^m,\ \nabla_x \psi(x,u) = 0,\ u \geqq 0\} \\[2mm] \psi(x,u) = \theta(x) + ug(x) \end{array} \right\rangle \tag{DP}$$

or equivalently

$$\left\langle \begin{array}{l} \theta(\hat{x}) + \hat{u}g(\hat{x}) = \max_{(x,u) \in Y} \theta(x) + ug(x) \\[2mm] (\hat{x},\hat{u}) \in Y = \{(x,u) \mid x \in X^0,\ u \in R^m, \\[1mm] \hspace{4cm} \nabla\theta(x) + u\nabla g(x) = 0,\ u \geqq 0\} \end{array} \right\rangle \tag{DP}$$

The duality results we are about to establish relate solutions $\bar{x}$ of MP and $(\hat{x},\hat{u})$ of DP to each other. They also relate the objective functions θ and ψ to each other. We begin by giving a weak and easy-to-derive duality theorem.

3   **Weak duality theorem [Wolfe 61]**

Let X^0 be open, and let θ and g be differentiable on X^0. Then

$$\left\langle \begin{array}{l} x^1 \in X \\[2mm] (x^2,u^2) \in Y \\[2mm] \theta \text{ and } g \text{ convex at } x^2 \end{array} \right\rangle \Rightarrow \theta(x^1) \geqq \psi(x^2,u^2)$$

where X and Y are defined in 1 and 2.

PROOF

$$\theta(x^1) \geqq \theta(x^2) + \nabla\theta(x^2)(x^1 - x^2) \qquad \text{(by } 6.1.1)$$

$$= \theta(x^2) - u^2\nabla g(x^2)(x^1 - x^2) \qquad \text{(since } \nabla\theta(x^2) = -u^2\nabla g(x^2))$$

$$\geqq \theta(x^2) + u^2[g(x^2) - g(x^1)] \qquad \text{(by } 6.1.1 \text{ and } u^2 \geqq 0)$$

$$\geqq \theta(x^2) + u^2 g(x^2) \qquad \text{(since } u^2 \geqq 0 \text{ and } g(x^1) \leqq 0)$$

$$= \psi(x^2, u^2) \quad \blacksquare$$

We derive now one of the more important duality theorems of nonlinear programming.

4   **Wolfe's duality theorem [Wolfe 61]**

Let X^0 be an open set in R^n, let θ and g be differentiable and convex on X^0, let $\bar{x}$ solve MP 1, and let g satisfy any one of the six constraint qualifications of Theorem 7.3.7. Then there exists a $\bar{u} \in R^m$ such that $(\bar{x}, \bar{u})$ solves DP 2 and

$$\theta(\bar{x}) = \psi(\bar{x}, \bar{u})$$

PROOF   [Huard 62]   By Theorem 7.3.7, there exists a $\bar{u} \in R^m$ such that $(\bar{x}, \bar{u})$ satisfies the Kuhn-Tucker conditions

$$\nabla\theta(\bar{x}) + \bar{u}\nabla g(\bar{x}) = 0$$

$$\bar{u}g(\bar{x}) = 0$$

$$g(\bar{x}) \leqq 0$$

$$\bar{u} \geqq 0$$

Hence

$$(\bar{x}, \bar{u}) \in Y = \{(x,u) \mid x \in X^0, u \in R^m, \nabla\theta(x) + u\nabla g(x) = 0, u \geqq 0\}$$

Now let (x,u) be an arbitrary element of the set Y. Then

$$\psi(\bar{x}, \bar{u}) - \psi(x,u) = \theta(\bar{x}) - \theta(x) + \bar{u}g(\bar{x}) - ug(x)$$

$$\geqq \nabla\theta(x)(\bar{x} - x) - ug(x)$$
$$\text{(by } 6.1.1 \text{ and } \bar{u}g(\bar{x}) = 0)$$

$$\geqq \nabla\theta(x)(\bar{x} - x) + u[-g(\bar{x}) + \nabla g(x)(\bar{x} - x)]$$
$$\text{(by } 6.1.1 \text{ and } u \geqq 0)$$

$$= [\nabla\theta(x) + u\nabla g(x)](\bar{x} - x) - ug(\bar{x})$$

$$= -ug(\bar{x}) \qquad \text{(since } \nabla\theta(x) + u\nabla g(x) = 0)$$

$$\geqq 0 \qquad \text{(since } u \geqq 0 \text{ and } g(\bar{x}) \leqq 0)$$

Hence

$$\psi(\bar{x},\bar{u}) = \max_{(x,u)\in Y} \psi(x,u) \qquad (\bar{x},\bar{u}) \in Y$$

Since

$$\bar{u}g(\bar{x}) = 0$$

$$\psi(\bar{x},\bar{u}) = \theta(\bar{x}) + \bar{u}g(\bar{x}) = \theta(\bar{x}) \quad \blacksquare$$

ALTERNATE PROOF   Again as above, by Theorem 7.3.7, there exists a $\bar{u} \in R^m$ such that $(\bar{x},\bar{u})$ satisfies the Kuhn-Tucker conditions.   Hence $(\bar{x},\bar{u}) \in Y$.   Now

$$\psi(\bar{x},\bar{u}) = \theta(\bar{x}) + \bar{u}g(\bar{x}) = \theta(\bar{x}) \geqq \psi(x,u) \qquad \text{for } (x,u) \in Y$$

where the last inequality follows from Theorem 3.   Hence $(\bar{x},\bar{u})$ solves DP 2. $\blacksquare$

REMARK   The constraint qualification in the above theorem is merely a sufficient condition for the theorem to hold.   It may hold without satisfying such a constraint qualification, as evidenced by the following primal problem.   Find $\bar{x} \in R^2$ such that

$$0 = \min_{x\in X} 0 \qquad (\bar{x}_1,\bar{x}_2) \in X = \left\{ x \left| \begin{array}{l} x \in R^2 \\ (x_1 + 1)^2 + (x_2)^2 \leqq \alpha \\ (x_1 - 1)^2 + (x_2)^2 \leqq \alpha \end{array} \right. \right\}$$

where α is some nonnegative number.   The dual to this problem is: Find $\hat{x} \in R^2$ and $\hat{u} \in R^2$ such that

$$\hat{u}_1[(\hat{x}_1 + 1)^2 + (\hat{x}_2)^2 - \alpha] + \hat{u}_2[(\hat{x}_1 - 1)^2 + (\hat{x}_2)^2 - \alpha]$$
$$= \max_{(x,u)\in Y} u_1[(x_1 + 1)^2 + (x_2)^2 - \alpha] + u_2[(x_1 - 1)^2 + (x_2)^2 - \alpha]$$

$$(\hat{x},\hat{u}) \in Y = \left\{ (x,u) \left| \begin{array}{l} x \in R^2, \ u \in R^2 \\ 2u_1(x_1 + 1) + 2u_2(x_1 - 1) = 0 \\ 2u_1x_2 + 2u_2x_2 = 0 \\ u_1 \geqq 0, \ u_2 \geqq 0 \end{array} \right. \right\}$$

When $\alpha = 1$, the primal problem has one feasible (and hence minimal) point, $\bar{x}_1 = \bar{x}_2 = 0$.   At this point none of the constraint qualifications are satisfied.   However it can be verified (after a little calculation) that $\bar{x} = 0$, $\bar{u} = 0$ also solves the dual problem with a maximum of zero.

Another important duality theorem is the converse of Theorem 4 above.   In order to obtain such a theorem, we have to modify the hypotheses of Theorem 4.   There are a number of such converse theorems [Hanson 61, Huard 62, Mangasarian 62, Mangasarian-Ponstein 65, Karamardian 67]. To cover all such theorems would be lengthy. Instead we shall confine ourselves to two converse theorems which hold under somewhat different assumptions and the new proofs of which are completely different from each other.

5   ### Strict converse duality theorem† [Mangasarian 62]

Let X^0 be an open set in R^n, let θ and g be differentiable and convex on X^0, let MP 1 have a solution $\bar{x}$, and let g satisfy any one of the six constraint qualifications of Theorem 7.3.7. If $(\hat{x},\hat{u})$ is a solution of DP 2, and if $\psi(x,\hat{u})$ is strictly convex at $\hat{x}$, then $\hat{x} = \bar{x}$, that is, $\hat{x}$ solves MP 1, and

$$\theta(\bar{x}) = \psi(\hat{x},\hat{u})$$

REMARK $\psi(x,\hat{u})$ is strictly convex at $\hat{x}$ if either θ is strictly convex at $\hat{x}$ or if for some i, $\hat{u}_i > 0$ and g_i is strictly convex at $\hat{x}$.

PROOF   We shall assume that $\hat{x} \neq \bar{x}$ and exhibit a contradiction. Since $\bar{x}$ is a solution of MP *1*, it follows from Theorem *4* above that there exists a $\bar{u} \in R^m$ such that $(\bar{x},\bar{u})$ solves DP *2*. Hence

$$\psi(\bar{x},\bar{u}) = \psi(\hat{x},\hat{u}) = \max_{(x,u)\in Y} \psi(x,u)$$

and

$$(\bar{x},\bar{u}) \in Y$$

Because $(\hat{x},\hat{u}) \in Y$, we have that $\nabla_x \psi(\hat{x},\hat{u}) = 0$. Hence by the strict convexity of $\psi(x,\hat{u})$ at $\hat{x}$ and *6.2.1* we have that

$$\psi(\bar{x},\hat{u}) - \psi(\hat{x},\hat{u}) > \nabla_x \psi(\hat{x},\hat{u})(\bar{x} - \hat{x}) = 0$$

It follows then that

$$\psi(\bar{x},\hat{u}) > \psi(\hat{x},\hat{u}) = \psi(\bar{x},\bar{u})$$

or that

$$\hat{u}g(\bar{x}) > \bar{u}g(\bar{x})$$

But from Theorem *7.3.7* we have that $\bar{u}g(\bar{x}) = 0$, hence

$$\hat{u}g(\bar{x}) > 0$$

† We have used the adjective *strict* to distinguish the above converse duality theorem from other theorems, such as Dorn's converse duality theorem [Dorn 60] (see *8.2.6* below) and theorem 5.6 of [Mangasarian-Ponstein 65], in which the solution of the primal problem $\bar{x}$ need not equal $\hat{x}$, where $(\hat{x},\hat{u})$ is the solution of the dual problem.

which contradicts the facts that $\hat{u} \geq 0$ and $g(\bar{x}) \leq 0$. Hence $\hat{x} = \bar{x}$. We also have that

$$\theta(\bar{x}) = \theta(\bar{x}) + \bar{u}g(\bar{x}) = \psi(\bar{x},\bar{u}) = \psi(\hat{x},\hat{u}) \quad \blacksquare$$

It should be remarked that the above theorem can be strengthened (see theorem 5.7 in [Mangasarian-Ponstein 65]) by dropping the assumptions that MP *1* has a solution and that g satisfies a constraint qualification, but retaining the strict convexity assumption on $\psi(x,\hat{u})$. The proof of this strengthened theorem is rather lengthy, and hence we content ourselves here with the above weaker version.

We derive now another strict converse duality theorem under different hypotheses from those of Theorem *5* above. We drop the assumptions that MP *1* has a solution, that g satisfies a constraint qualification, and that $\psi(x,\hat{u})$ is strictly convex at $\hat{x}$. We add the assumptions that $\psi(x,\hat{u})$ is twice continuously differentiable at $\hat{x}$ and that the Hessian matrix $\nabla_x{}^2\psi(x,\hat{u})$ is nonsingular at $\hat{x}$.

6   **The Hanson-Huard strict converse duality theorem [Hanson 61, Huard 62]**

Let X^0 be an open set in R^n, and let θ and g be differentiable on X^0. Let $(\hat{x},\hat{u})$ be a solution of DP 2, and let θ and g be convex at $\hat{x}$. If either (i) *[Huard 62] $\psi(x,\hat{u})$ is twice continuously differentiable at $\hat{x}$ and the $n \times n$ Hessian matrix $\nabla_x{}^2\psi(\hat{x},\hat{u})$ is nonsingular, or* (ii) *[Hanson 61] there exists an open set $\Lambda \subset R^m$ containing $\hat{u}$ and an n-dimensional differentiable vector function $e(u)$ on Λ such that*

$$\hat{x} = e(\hat{u})$$

and

$$\left\langle \begin{array}{c} e(u) \in X^0 \\[2mm] \nabla_x\psi(x,u)\Big|_{x=e(u)} = 0 \end{array} \right\rangle \quad \textit{for } u \in \Lambda$$

then $\hat{x}$ solves MP 1, and

$$\theta(\hat{x}) = \psi(\hat{x},\hat{u})$$

PROOF   Since $\nabla_x\psi(\hat{x},\hat{u}) = 0$, we have that assumption (i) above implies, by the implicit function theorem *D.3.1*, assumption (ii) above. (Assumption (ii) does not necessarily imply assumption (i).) We establish the theorem now under assumption (ii). From (ii) we have that

$$(\hat{x},\hat{u}) = [e(\hat{u}),\hat{u}] \in \{[e(u),u] \mid u \in \Lambda,\ u \geq 0\} \subset Y$$

and since

$$\psi[e(\hat{u}),\hat{u}] = \psi(\hat{x},\hat{u}) = \max_{(x,u)\in Y} \psi(x,u)$$

we have then that

$$\psi[e(\hat{u}),\hat{u}] = \max_{u\in\Lambda} \{\psi[e(u),u] \mid u \geq 0\}$$

Since Λ is open and the constraint $u \geq 0$ is linear, we have by Theorem 7.3.7 (or more conveniently by 7.3.10) and the chain rule D.1.6 that

$$\nabla_x\psi(x,\hat{u})\Big|_{x=e(\hat{u})} \nabla_u e(\hat{u}) + \nabla_u\psi(\hat{x},\hat{u})\Big|_{\hat{x}=e(\hat{u})} \leq 0$$

$$\hat{u}\left[\nabla_x\psi(x,\hat{u})\Big|_{x=e(\hat{u})} \nabla_u e(\hat{u}) + \nabla_u\psi(\hat{x},\hat{u})\Big|_{\hat{x}=e(\hat{u})}\right] = 0$$

Since $\hat{u} \in \Lambda$, since $\nabla_x\psi(x,u)\Big|_{x=e(u)} = 0$ for $u \in \Lambda$, and since

$$\psi(x,u) = \theta(x) + ug(x)$$

the last two relations give

$$g(\hat{x}) \leq 0$$

$$\hat{u}g(\hat{x}) = 0$$

But since $(\hat{x},\hat{u}) \in Y$, we also have that

$$\nabla\theta(\hat{x}) + \hat{u}\nabla g(\hat{x}) = 0$$

$$\hat{u} \geq 0$$

The last four relations and Theorem 7.2.1 imply that $\hat{x}$ is a solution of MP 1. We also have that

$$\theta(\hat{x}) = \theta(\hat{x}) + \hat{u}g(\hat{x}) = \psi(\hat{x},\hat{u}) \quad \blacksquare$$

We give now a theorem which shows when the dual problem has an objective function, unbounded from above, on the set of dual feasible points Y. This is a generalization of a theorem of Wolfe [Wolfe 61, theorem 3].

7   **Unbounded dual theorem**

Let X^0 be an open set in R^n, and let θ and g be differentiable on X^0. If there exists a dual feasible point (x^1,u^1) such that

$$\langle g(x^1) + \nabla g(x^1)z \leq 0 \text{ has no solution } z \in R^n\rangle$$

then the dual problem has an unbounded objective function (from above) on the set of dual feasible points Y.

PROOF   By Theorem *2.4.10* of the alternative there exists a $u^2 \in R^m$ such that

$$u^2 \nabla g(x^1) = 0$$

$$u^2 \geqq 0$$

$$u^2 g(x^1) = 1$$

Since (x^1, u^1) is dual feasible, then

$$\nabla \theta(x^1) + u^1 \nabla g(x^1) = 0$$

$$u^1 \geqq 0$$

Consider now the point $(x^1, u^1 + \tau u^2)$ for any $\tau > 0$.   We have that

$$\nabla \theta(x^1) + (u^1 + \tau u^2) \nabla g(x^1) = [\nabla \theta(x^1) + u^1 \nabla g(x^1)] + \tau u^2 \nabla g(x^1) = 0$$

$$u^1 + \tau u^2 \geqq 0$$

and hence $(x^1, u^1 + \tau u^2)$ is dual feasible for any $\tau > 0$.   Since $u^2 g(x^1) = 1$, we have that the dual objective function

$$\psi(x^1, u^1 + \tau u^2) = \theta(x^1) + (u^1 + \tau u^2) g(x^1) = \theta(x^1) + u^1 g(x^1) + \tau$$

tends to ∞ as τ tends to ∞.   ∎

REMARK   The above theorem does not apply if g is convex at x^1 and the primal problem has a feasible point, x^*, say.   For if we define $z = x^* - x^1$, then we have that

$$0 \geqq g(x^*) = g(x^1 + z) \geqq g(x^1) + \nabla g(x^1)z$$

and hence $g(x^1) + \nabla g(x^1)z \leqq 0$ has a solution $z = x^* - x^1$.

EXAMPLE   Consider the example in the Remark following Theorem *4* above, and let $\alpha = \frac{1}{4}$.   Obviously then the primal problem has no feasible point.   Consider now the dual feasible point $x_1^1 = x_2^1 = 0$, $u_1^1 = u_2^1 = 1$.   We have then that

$$g(x^1) + \nabla g(x^1)z = \begin{pmatrix} \frac{3}{4} \\ \frac{3}{4} \end{pmatrix} + \begin{pmatrix} 2 & 0 \\ -2 & 0 \end{pmatrix} z \leqq 0$$

has no solution $z \in R^n$, and hence by Theorem *7* above the dual objective function is unbounded from above on the dual feasible region Y.   This can be easily seen if we let $x_1 = x_2 = 0$ and let $u_1 = u_2$ tend to ∞.   Then $\psi(x,u)$ tends to ∞.

8   Corollary (unbounded dual)

Let $X^0 = R^n$, and let θ and g be differentiable on R^n. Let the dual problem DP 2 have a feasible point (x^1, u^1), let g be concave† at x^1 or linear on R^n, and let the primal problem MP 1 have no feasible point. Then the dual problem has an unbounded objective function $\psi(x, u)$ (from above) on the set of dual feasible points Y.

PROOF   We claim that

$\langle g(x^1) + \nabla g(x^1)z \leqq 0$ has no solution $z \in R^n \rangle$

For if it did have a solution $z \in R^n$, then $x^* = x^1 + z$ would satisfy

$g(x^*) = g(x^1 + z) \leqq g(x^1) + \nabla g(x^1)z \leqq 0$

where the next-to-last inequality follows from 6.1.1.   Hence

$x^* \in X = \{x \mid x \in R^n, g(x) \leqq 0\}$

which contradicts the assumption that X is empty.   So,

$g(x^1) + \nabla g(x^1)z \leqq 0$

has no solution z in R^n, and by Theorem 7 the dual problem has an unbounded objective function (from above) on the set Y of dual feasible points. ∎

The case when g is linear in the above corollary is theorem 3 of [Wolfe 61].

We finally give a theorem which tells us when the primal problem has no local or global minimum.

9   Theorem (no primal minimum)

Let X^0 be an open set in R^n, let θ and g be differentiable and concave‡ on X^0, and let $X \neq \emptyset$.   If the dual problem DP 2 has no feasible point, then neither the primal problem MP 1 nor the local minimization problem LMP 7.1.2 has a solution.

PROOF   Because the dual problem has no feasible point, we have that

$\left\langle \begin{array}{c} u\nabla g(x) = -\nabla \theta(x) \\ u \geqq 0 \end{array} \right\rangle$ has no solution $u \in R^m$ for each $x \in X^0$

† Note that if g were concave on R^n, the set X would still not be convex in general, unless g were linear.

‡ Note that the concavity of θ and g does not, in general, produce a primal convex programming problem (unless θ and g are linear).

Hence by Farkas' theorem *2.4.6*

$$\left\langle \begin{matrix} \nabla g(x)z \leq 0 \\ \\ \nabla \theta(x)z < 0 \end{matrix} \right\rangle \begin{matrix} \text{has a solution } z \in R^n, \|z\| < \delta, \text{ for each } x \in X^0 \text{ and any} \\ \delta > 0 \end{matrix}$$

Let $x \in X$. Then by *6.1.1* and the last two inequalities we have that

$$g(x + z) \leq g(x) + \nabla g(x)z \leq 0$$

$$\theta(x + z) \leq \theta(x) + \nabla \theta(x)z < \theta(x)$$

Hence by choosing δ small enough, $x + z \in B_{\bar{\delta}}(x) \cap X$ for any $\bar{\delta} > 0$ and x cannot be a solution of LMP *7.1.2* and, hence, not a solution of MP *1*. ∎

We mentioned earlier that there is an extensive literature on the duality theory of nonlinear programming. The presentation we have followed here resembles that given in [Dorn 60, Wolfe 61, Hanson 61, Huard 62, Mangasarian 62]. There are approaches to duality which use the conjugate function concept, which we have not touched upon here [Fenchel 53, Berge-Ghouila Houri 65, Dieter 65a, Rockafellar 63, 64, 67a, 67b, 67c, 69, Whinston 65, 67]. There are also approaches that use a minmax theorem [Stoer 63, 64, Mangasarian-Ponstein 65, Karamardian 67]. There is also the theory of *symmetric* dual nonlinear programs [Dorn 61, Cottle 63, Dantzig et al. 65, Mond 65, Mond-Cottle 66, Stoer 64]. Duality relations relaxing convexity conditions have appeared in [Mangasarian 65, Karamardian 67, Rissanen 67, Jacob-Rossi 69] (see also Chap. 10). Duality in spaces more general than R^n has also appeared in [Hurwicz 58, Kretschmar 61, Rubinstein 63, Rockafellar 63, Brondsted 64, Moreau 65, Tyndall 65, 67, Dieter 65b, 66, Levinson 66, Larsen-Polak 66, Gol'stein 67, Ritter 67, Varaiya 67, Hanson-Mond 67, Van Slyke-Wets 67, Hanson 68].

10   **Problem**

Let X^0 be an open set in $R^{n_1} \times R^{n_2}$, let θ and g be respectively a differentiable numerical function and a differentiable m-dimensional vector function on X^0. Show that the dual to the following primal minimization problem

$$\theta(\bar{x},\bar{y}) = \min_{(x,y \in X)} \theta(x,y) \qquad (\bar{x},\bar{y}) \in X = \{(x,y) \mid (x,y) \in X^0, g(x,y) \leq 0,$$

$$y \geq 0\}$$

is the following maximization problem

$$
\left|
\begin{array}{l}
\psi(\hat{x},\hat{y},\hat{u}) - \hat{y}\nabla_y\psi(\hat{x},\hat{y},\hat{u}) = \max_{(x,y,u)\in Y} \psi(x,y,u) - y\nabla_y\psi(x,y,u) \\[2mm]
(\hat{x},\hat{y},\hat{u}) \in Y = \left\{ (x,y,u) \left|
\begin{array}{l}
(x,y) \in X^0, u \in R^m \\
\nabla_x\psi(x,y,u) = 0 \\
\nabla_y\psi(x,y,u) \geqq 0 \\
u \geqq 0
\end{array}
\right. \right\} \\[6mm]
\psi(x,y,u) = \theta(x,y) + ug(x,y)
\end{array}
\right.
$$

State and prove the duality relations that hold for these dual problems and the conditions under which they hold.

2. Duality in quadratic programming

We consider in this section programming problems with a quadratic objective function and linear constraints. Let b be an n-vector, c an m-vector, C a symmetric† $n \times n$ matrix, and A an $m \times n$ matrix. We consider the following primal problem.

1

The (primal) quadratic minimization problem (QMP)

Find an $\bar{x}$, if it exists, such that

$$
\tfrac{1}{2}\bar{x}C\bar{x} - b\bar{x} = \min_{x\in X} \tfrac{1}{2}xCx - bx \qquad \bar{x} \in X = \{x \mid x \in R^n, Ax \leqq c\}
$$

$$(\text{QMP})$$

According to 8.1.2 the dual to the above problem is given by

$$
\tfrac{1}{2}\hat{x}C\hat{x} - b\hat{x} + \hat{u}(A\hat{x} - \hat{c}) = \max_{(x,u)\in Y} \tfrac{1}{2}xCx - bx + u(Ax - c)
$$

$$
(\hat{x},\hat{u}) \in Y = \left\{ (x,u) \left|
\begin{array}{l}
x \in R^n, u \in R^m \\
Cx - b + A'u = 0 \\
u \geqq 0
\end{array}
\right. \right\}
$$

The constraint relation $Cx - b + A'u = 0$ implies that

$$
xCx - bx + uAx = 0
$$

Using this equation in the objective function, the dual problem becomes the following.

† If C is not symmetric, we replace C by $(C + C')/2$ in QMP 1 because $xCx = x[(C + C')/2]x$.

2　　　The quadratic dual (maximization) problem (QDP)

Find an $\hat{x} \in R^n$ and a $\hat{u} \in R^m$, if they exist, such that

$$\left\langle \begin{array}{l} -\tfrac{1}{2}\hat{x}C\hat{x} - c\hat{u} = \max_{(x,u)\in Y} (-\tfrac{1}{2}xCx - cu) \\[2mm] (\hat{x},\hat{u}) \in Y = \left\{(x,u) \left| \begin{array}{l} x \in R^n,\ u \in R^m \\ Cx + A'u = b \\ u \geq 0 \end{array} \right.\right\} \end{array} \right\rangle \qquad (\text{QDP})$$

A whole group of duality theorems now follows from the theorems of the previous section.

3　　　Weak duality theorem

Let C be positive semidefinite (that is, $xCx \geq 0$ for all x in R^n). Then

$$\left\langle \begin{array}{l} x^1 \in X \\[1mm] (x^2,u^2) \in Y \end{array} \right\rangle \Rightarrow \tfrac{1}{2}x^1Cx^1 - bx^1 \geq -\tfrac{1}{2}x^2Cx^2 - cu^2$$

PROOF This theorem follows from Theorem *8.1.3* by observing that $\tfrac{1}{2}xCx - bx$ is convex on R^n if the matrix C is positive semidefinite (see Theorem *6.3.2*). ∎

4　　　Dorn's duality theorem [Dorn 60]

Let C be positive semidefinite. If $\bar{x}$ solves QMP 1, then $\bar{x}$ and some $\bar{u} \in R^m$ solve QDP 2, and the two extrema are equal.

PROOF This theorem follows from Theorem *8.1.4* if we make the same observation as in the proof of Theorem *3* above. ∎

REMARK The dual quadratic programs *1* and *2* possess a nice feature not shared by the dual nonlinear programs *8.1.1* and *8.1.2*. If C is positive semidefinite, then the objective function of QMP *1* is convex on R^n, and the objective function of QDP *2* is concave on $R^n \times R^m$.

5　　　Strict converse duality theorem

Let C be positive definite (that is, $xCx > 0$ for all nonzero x in R^n). If $(\hat{x},\hat{u})$ solves QDP 2, then $\hat{x}$ solves QMP 1, and the two extrema are equal.

PROOF This theorem follows from Theorem *8.1.6* by observing that if C is positive definite, then $\tfrac{1}{2}xCx - bx$ is strictly convex (and hence convex) on R^n (see Theorem *6.4.2*), and the Hessian matrix $\nabla_x^2[\tfrac{1}{2}xCx - bx + \hat{u}(Ax - c)]$, which is equal to C, is nonsingular (for if it were singular, then $Cx = 0$ for some $x \neq 0$, and $xCx = 0$ for some $x \neq 0$, which contradicts the positive definite assumption). ∎

6   ## Dorn's converse duality theorem [Dorn 60]

Let C be positive semidefinite. If $(\hat{x}, \hat{u})$ solves QDP 2, then some $\bar{x} \in R^n$ (not necessarily equal to $\hat{x}$), satisfying $C(\bar{x} - \hat{x}) = 0$, solves QMP 1, and the two extrema are equal.

REMARK We have not established here the counterpart of this theorem for the general nonlinear case. Such a theorem exists [Mangasarian-Ponstein 65, theorem 5.6]. Note the difference between *5* and *6* above. In *5* the *same* $\hat{x}$ appears in the solutions of *1* and *2*. In *6* we merely have that $C(\bar{x} - \hat{x}) = 0$.

PROOF By Theorem *7.3.7* (or more conveniently by *7.3.12*) and the linearity of the constraints $Cx + A'u = b$, $u \geq 0$, there exists a $\hat{v} \in R^n$ such that $(\hat{x}, \hat{u}, \hat{v})$ satisfies the following Kuhn-Tucker conditions

$$-C\hat{x} + C\hat{v} = 0$$

$$C\hat{x} + A'\hat{u} = b$$

$$-c + A\hat{v} \leq 0$$

$$-c\hat{u} + \hat{u}A\hat{v} = 0$$

$$\hat{u} \geq 0$$

Substitution of the first relation into the second one gives

$$C\hat{v} + A'\hat{u} = b$$

The last four relations, the assumption that C is positive semidefinite, and Theorems *6.3.2* and *7.2.3* imply that $\hat{v}$ solves QMP *1*. Hence $\bar{x} = \hat{v}$ solves QMP *1*, and $C\bar{x} = C\hat{v} = C\hat{x}$. Now we show that the two extrema are equal.

$$
\begin{aligned}
(-b\bar{x} + \tfrac{1}{2}\bar{x}C\bar{x}) &- (-c\hat{u} - \tfrac{1}{2}\hat{x}C\hat{x}) \\
&= -b\hat{v} + \tfrac{1}{2}\hat{v}C\hat{v} + c\hat{u} + \tfrac{1}{2}\hat{x}C\hat{x} \qquad \text{(since } \bar{x} = \hat{v}) \\
&= -b\hat{v} + \hat{v}C\hat{x} + c\hat{u} \qquad \text{(since } C\hat{x} = C\hat{v} \text{ and } C = C') \\
&= -\hat{u}A\hat{v} + c\hat{u} \qquad \text{(since } C\hat{x} + A'\hat{u} = b) \\
&= 0 \qquad \text{(since } -c\hat{u} + \hat{u}A\hat{v} = 0) \quad ∎
\end{aligned}
$$

7   Unbounded dual theorem

           Let $Y \neq \emptyset$.   Then

$\langle X = \emptyset \rangle \Rightarrow \langle QDP\ 2\ has\ an\ unbounded\ objective\ function\ from\ above\ on\ Y \rangle$

PROOF   Follows from Corollary 8.1.8.   ∎

8   Theorem (no primal minimum)

           Let C be negative semidefinite, and let $X \neq \emptyset$.   Then

$\langle Y = \emptyset \rangle \Rightarrow \langle QMP\ 1\ has\ no\ solution \rangle$

PROOF   Follows from Theorem 8.1.9.   ∎

9   Problem

           Let b, a, c, and d be given vectors in R^n, R^ℓ, R^m, and R^k, respectively, and let A, D, B, E, C, and F be given $m \times n$, $m \times \ell$, $k \times n$, $k \times \ell$, $n \times n$ (symmetric), and $\ell \times \ell$ (symmetric) matrices, respectively.   Show that the following are dual quadratic problems.

$$\left|\begin{array}{l} -b\bar{x} + \tfrac{1}{2}\bar{x}C\bar{x} - a\bar{y} + \tfrac{1}{2}\bar{y}F\bar{y} \\ \qquad = \min_{(x,y)\in X} \left(-bx + \tfrac{1}{2}xCx - ay + \tfrac{1}{2}yFy\right) \\[2mm] (\bar{x},\bar{y}) \in X = \left\{(x,y)\ \middle|\ \begin{array}{l} x \in R^n,\ y \in R^\ell \\ Ax + Dy \leq c \\ Bx + Ey = d \\ y \geqq 0 \end{array}\right\} \end{array}\right\rangle \quad \text{(QMP)}$$

$$\left|\begin{array}{l} -\tfrac{1}{2}\hat{x}C\hat{x} - \tfrac{1}{2}\hat{y}F\hat{y} - c\hat{u} - d\hat{v} \\ \qquad = \max_{(x,y,u,v)\in Y} \left(-\tfrac{1}{2}xCx - \tfrac{1}{2}yFy - cu - dv\right) \\[2mm] (\hat{x},\hat{y},\hat{u},\hat{v}) \in Y = \left\{(x,y,u,v)\ \middle|\ \begin{array}{l} x \in R^n,\ y \in R^\ell,\ u \in R^m,\ v \in R^k \\ Cx + A'u + B'v = b \\ Fy + D'u + E'v \geqq a \\ u \geqq 0 \end{array}\right\} \end{array}\right\rangle \quad \text{(QDP)}$$

3.   Duality in linear programming

We outline briefly here the main duality results of linear programming and show how they follow from the results we have established.

           By deleting the matrix C from the dual quadratic problems 8.2.1 and 8.2.2, we obtain the following dual linear programs.

1   The (primal) linear minimization problem (LMP)

Find an $\bar{x}$, if it exists, such that

$$-b\bar{x} = \min_{x \in X} (-bx) \qquad \bar{x} \in X = \{x \mid x \in R^n, \, Ax \leqq c\} \qquad \text{(LMP)}$$

2   The dual linear (maximization) problem (LDP)

Find a $\bar{u}$, if it exists, such that

$$-c\bar{u} = \max_{u \in Y} (-cu) \qquad \bar{u} \in Y = \{u \mid u \in R^m, \, A'u = b, \, u \geqq 0\} \quad \text{(LDP)}$$

Note that the dual problem contains only the variable u. The nonlinear and quadratic dual problems *8.1.2* and *8.2.2* contain both the variables x and u. The dual linear program has the unique feature that it does not contain the primal variable x.

We combine now all the fundamental duality results of linear programming in the following theorem.

3   Duality theorem of linear programming [Gale et al. 51]

(i) $\left\langle \begin{array}{c} x \in X \\ u \in Y \end{array} \right\rangle \Rightarrow -bx \geqq -cu$

(ii) $\langle \bar{x} \text{ solves } LMP \; 1 \rangle \Leftrightarrow \langle \bar{u} \text{ solves } LDP \; 2 \rangle \Rightarrow -b\bar{x} = -c\bar{u}$

(iii)   Let $Y \neq \emptyset$. Then

$\langle X = \emptyset \rangle \Leftrightarrow \langle -cu \text{ is unbounded from above on } Y \rangle$

(iv)   Let $X \neq \emptyset$. Then

$\langle Y = \emptyset \rangle \Leftrightarrow \langle -bx \text{ is unbounded from below on } X \rangle$

(v) $\langle X \neq \emptyset \text{ and } Y \neq \emptyset \rangle \Leftrightarrow \left\langle \begin{array}{c} \exists \bar{x} \text{ that solves } LMP \; 1 \\ \text{and} \\ \exists \bar{u} \text{ that solves } LDP \; 2 \end{array} \right\rangle$

REMARK   Both of the sets X and Y may be empty, in which case none of the above results hold. For example take $A = 0$, $c = -1$, $b = -1$.

PROOF   (i)   Follows from *8.2.3*.

(ii)   Follows from *8.2.4* and *8.2.6*.

(iii)   The forward implication follows from *8.2.7*. The backward implication is equivalent logically to

$\langle X \neq \emptyset \rangle \Rightarrow \langle -cu \text{ is bounded from above on } Y \rangle$

which follows from part (i) of this theorem.

(iv) $(\Rightarrow)$

$\langle Y = \emptyset \rangle \Rightarrow \langle A'u = b,\ u \geq 0$ has no solution $u \in R^m \rangle$

$\qquad \Rightarrow \langle \exists x^1 : bx^1 = 1,\ Ax^1 \leq 0 \rangle \qquad\qquad\qquad \text{(by 2.4.6)}$

$\qquad \Rightarrow \left\langle \begin{array}{l} A(\lambda x^1 + x^2) \leq c \\ -b(\lambda x^1 + x^2) = -\lambda + bx^2 \\ \text{for any } x^2 \in X \text{ and any } \lambda \geq 0 \end{array} \right\rangle$

$\qquad \Rightarrow \left\langle \begin{array}{l} -bx \text{ is unbounded from below on } X \\ \qquad\qquad\qquad (\text{let } \lambda \to \infty) \end{array} \right\rangle$

$(\Leftarrow)$ The backward implication is equivalent to

$\langle Y \neq \emptyset \rangle \Rightarrow \langle -bx \text{ is bounded from below on } X \rangle$

which follows from part (i) of this theorem.

(v) The backward implication is trivial because $\bar{x} \in X$ and $\bar{u} \in Y$. We prove now the forward implication. Because of (ii) we need only prove that

$\langle X \neq \emptyset \text{ and } Y \neq \emptyset \rangle \Rightarrow \langle \exists \bar{x} \text{ that solves LMP } 1 \rangle$

Since $X \neq \emptyset$, we have that $Ax \leq c$ has a solution x, which implies that $A'v = 0,\ v \geq 0,\ cv < 0$ has no solution v (for if it did have a solution v, then $0 = vAx \leq cv$, which contradicts $cv < 0$). Hence

4    $\langle A'v = 0,\ v \geq 0 \rangle \Rightarrow cv \geq 0$

Similarly, since $Y \neq \emptyset$, $A'u = b,\ u \geq 0$ has a solution u, which implies that $Ay \leq 0,\ by > 0$ has no solution y (for if it did have a solution y, then $0 \geq uAy = by$, which contradicts $by > 0$). Hence

5    $Ay \leq 0 \Rightarrow by \leq 0$

We will now show that if LMP 1 has no solution, then a contradiction ensues.

6    $\langle \text{LMP } 1 \text{ has no solution} \rangle$

$\qquad \Rightarrow \left\langle \begin{array}{c} Ax \leq c \\ A'u = b \\ u \geq 0 \\ -bx + cu = 0 \end{array} \right\rangle \quad \text{has no solution } (x,u) \qquad \text{(by 2.2.2)}$

$$\Rightarrow \left\langle \begin{array}{l} Ax \leqq c \\ A'u = b \\ u \geqq 0 \\ -bx + cu \leqq 0 \end{array} \right\rangle \quad \begin{array}{l} \text{has no solution } (x,u) \qquad \text{(since the first} \\ \text{3 relations imply that } -bx + cu \geqq 0) \end{array}$$

$$\Rightarrow \left\langle \begin{array}{l} -\zeta < 0 \\ Ax \qquad - c\zeta \leqq 0 \\ A'u - b\zeta \leqq 0 \\ -A'u + b\zeta \leqq 0 \\ -u \qquad \leqq 0 \\ -bx + cu \qquad \leqq 0 \end{array} \right\rangle \quad \begin{array}{l} \text{has no solution } (x,u,\zeta),\; x \in R^n, \\ u \in R^m,\; \zeta \in R \end{array}$$

$$\Rightarrow \left\langle \begin{array}{l} A't - b\sigma = 0 \\ Ap - Aq - r + c\sigma = 0 \\ -\eta - ct - bp + bq = 0 \\ \eta > 0,\; (t,p,q,r,\sigma) \geqq 0 \end{array} \right\rangle \quad \begin{array}{l} \text{has a solution } (\eta,t,p,q,r,\sigma),\; \eta \in R, \\ t \in R^m,\; p \in R^n,\; q \in R^n, \\ r \in R^m,\; \sigma \in R \text{ (by 2.4.2)} \end{array}$$

$$\Rightarrow \left\langle \begin{array}{l} A't = b\sigma \\ -Az + c\sigma = r \\ bz - ct = \eta \\ \eta > 0,\; (t,r,\sigma) \geqq 0 \end{array} \right\rangle \quad \begin{array}{l} \text{has a solution } (\eta,t,z,r,\sigma) \\ \text{(follows by setting } z = q - p) \end{array}$$

$$\Rightarrow \left\langle \begin{array}{l} A't = b\sigma \\ Az \leqq c\sigma \\ bz - ct > 0 \\ (t,\sigma) \geqq 0 \end{array} \right\rangle \quad \text{has a solution } (t,z,\sigma)$$

Now, either $\sigma = 0$ or $\sigma > 0$. If $\sigma = 0$, then $Az \leqq 0$, $A't = 0$, $t \geqq 0$, which implies (by making use of 4 and 5 above) that $bz \leqq 0$, $ct \geqq 0$, and $bz - ct \leqq 0$. This contradicts $bz - ct > 0$ above. If $\sigma > 0$, we have, upon normalization, that $A't = b$, $Az \leqq c$, $-bz + ct < 0$, $t \geqq 0$, which is a contradiction since $A't = b$, $Az \leqq c$, and $t \geqq 0$ imply that $-bz + ct \geqq 0$. So in either case of $\sigma = 0$ or $\sigma > 0$ we arrive at a contradiction. Hence LMP 1 has a solution. ∎

7   **Problem**

Consider the primal linear minimization problem LMP 1. Show that if $X \neq \emptyset$, and if, for all x in X, $-bx \geq \alpha$ for some real number α, then LMP 1 has a solution $\bar{x}$. (Hint: Use the facts that $X \neq \emptyset$ and that $-bx < \alpha$ has no solution x in X to show that there exists a dual feasible point, that is, a $u \in Y$; then use Theorem 3, part (v).)

8   **Remark**

It is not true in general that if a linear function is bounded from below on a closed convex set, then it achieves its infimum on that set. For example, $x_2 \geq 0$ on the closed convex set $\{x \mid x \in R^2, (2)^{-x_1} - x_2 \leq 0\}$; however x_2 does not attain its infimum of zero on the set.† Problem 7 above shows that if a linear function is bounded from below on a *polytope*, then it achieves its infimum on the polytope.

9   **Problem**

Show that the following are dual linear programs:

$$
\left|
\begin{array}{l}
-b\bar{x} - a\bar{y} = \min_{(x,y) \in X} \; (-bx - ay) \\[2mm]
(\bar{x}, \bar{y}) \in X = \left\{(x,y) \; \middle| \;
\begin{array}{l}
x \in R^n, \, y \in R^\ell \\[1mm]
Ax + Dy \leq c \\[1mm]
Bx + Ey = d \\[1mm]
y \geq 0
\end{array}
\right\}
\end{array}
\right. \qquad \text{(LMP)}
$$

$$
\left|
\begin{array}{l}
-c\bar{u} - d\bar{v} = \max_{(u,v) \in Y} \; (-cu - dv) \\[2mm]
(\bar{u}, \bar{v}) \in Y = \left\{(u,v) \; \middle| \;
\begin{array}{l}
u \in R^m, \, v \in R^k \\[1mm]
A'u + B'v = b \\[1mm]
D'u + E'v \geq a \\[1mm]
u \geq 0
\end{array}
\right\}
\end{array}
\right. \qquad \text{(LDP)}
$$

where b, a, c, and d are given vectors in R^n, R^ℓ, R^m, and R^k, respectively and A, D, B, and E are given $m \times n$, $m \times \ell$, $k \times n$, and $k \times \ell$ matrices, respectively.

† I am indebted to my colleague James W. Daniel for this example.

Chapter Nine

Generalizations of Convex Functions: Quasiconvex, Strictly Quasiconvex and Pseudoconvex Functions

Beginning with Chap. 4, we have continually used the concepts of convex and concave functions in deriving optimality conditions and duality relations. Since not all properties of convex and concave functions are needed in establishing some of the previous results, a more general type of function will also work. For example, some results need only that the set $\Lambda_\alpha = \{x \mid x \in \Gamma, \theta(x) \leqq \alpha\}$ be convex, where Γ is a convex set in R^n, θ is a numerical function defined on Γ, and α is any real number. Now if θ is a convex function on Γ, the convexity of Λ_α is assured by Theorem *4.1.10*. However, θ need not be convex in order that Λ_α be convex. A function θ which is *quasiconvex* on Γ has this property [Nikaidô 54]. Another property of differentiable convex functions that was used in previous results was this: If $\nabla\theta(\bar{x})(x - \bar{x}) \geqq 0$, then $\theta(x) \geqq \theta(\bar{x})$. This property follows from *6.1.1*, and has the obvious consequence that if $\nabla\theta(\bar{x}) = 0$, then $\theta(x) \geqq \theta(\bar{x})$. Not only convex functions have this property. *Pseudoconvex* functions [Tuy 64, Mangasarian 65], to be introduced in this chapter, also have this property. By using only the properties of convex functions that are needed in establishing some of the previous results, these results are extended to a larger class of functions.

1. Quasiconvex and quasiconcave functions

Quasiconvex function

A numerical function θ defined on a set $\Gamma \subset R^n$ is said to be *quasiconvex at* $\bar{x} \in \Gamma$ (with respect to Γ) if for each $x \in \Gamma$ such that $\theta(x) \leq \theta(\bar{x})$, the function θ assumes a value no larger than $\theta(\bar{x})$ on each point in the intersection of the closed line segment $[\bar{x},x]$ and Γ, or equivalently

$$\left.\begin{array}{c} x \in \Gamma \\ \theta(x) \leq \theta(\bar{x}) \\ 0 \leq \lambda \leq 1 \\ (1-\lambda)\bar{x} + \lambda x \in \Gamma \end{array}\right\} \Rightarrow \theta[(1-\lambda)\bar{x} + \lambda x] \leq \theta(\bar{x})$$

θ is said to be *quasiconvex on* Γ if it is quasiconvex at each $x \in \Gamma$.

Note that this definition of a quasiconvex function is slightly more general than the customary definition in the literature [Nikaidô 54, Berge-Ghouila Houri 65, Martos 65] in that (i) we define quasiconvexity at a point first and then on a set, and (ii) we do not require Γ to be convex. This generalization will allow us to handle a somewhat wider class of problems. It follows from the above definition that a numerical function θ defined on a convex set Γ is quasiconvex on Γ if and only if

$$\left.\begin{array}{c} x^1,x^2 \in \Gamma \\ \theta(x^2) \leq \theta(x^1) \\ 0 \leq \lambda \leq 1 \end{array}\right\} \Rightarrow \theta[(1-\lambda)x^1 + \lambda x^2] \leq \theta(x^1)$$

2   Quasiconcave function

A numerical function θ defined on a set $\Gamma \subset R^n$ is said to be *quasiconcave at* $\bar{x} \in \Gamma$ (with respect to Γ) if for each $x \in \Gamma$ such that $\theta(x) \geq \theta(\bar{x})$, the function θ assumes a value no smaller than $\theta(\bar{x})$ on each point on the intersection of the closed line segment $[\bar{x},x]$ and Γ, or equivalently

$$\left.\begin{array}{c} x \in \Gamma \\ \theta(\bar{x}) \leq \theta(x) \\ 0 \leq \lambda \leq 1 \\ (1-\lambda)\bar{x} + \lambda x \in \Gamma \end{array}\right\} \Rightarrow \theta(\bar{x}) \leq \theta[(1-\lambda)\bar{x} + \lambda x]$$

θ is said to be *quasiconcave on* Γ if it is quasiconcave at each $x \in \Gamma$.

Obviously θ is quasiconcave at $\bar{x}$ (on Γ) if and only if $-\theta$ is quasiconvex at $\bar{x}$ (on Γ). Results obtained for quasiconvex functions can be changed into results for quasiconcave functions by the appropriate multiplication by -1, and vice versa.

It follows from *2* that a numerical function θ defined on a convex set Γ is quasiconcave on Γ if and only if

$$\left.\begin{array}{l} x^1, x^2 \in \Gamma \\ \theta(x^1) \leq \theta(x^2) \\ 0 \leq \lambda \leq 1 \end{array}\right\} \Rightarrow \theta(x^1) \leq \theta\left[(1-\lambda)x^1 + \lambda x^2\right]$$

Figure *9.1.1* depicts a quasiconvex and a quasiconcave function on R.

3   **Theorem**

Let θ be a numerical function defined on a convex set $\Gamma \subset R^n$, let

$$\Lambda_\alpha = \{x \mid x \in \Gamma, \theta(x) \leq \alpha\}$$

and let

$$\Omega_\alpha = \{x \mid x \in \Gamma, \theta(x) \geq \alpha\}$$

Then

⟨θ quasiconvex on Γ⟩ ⟺ ⟨Λ_α is convex for each $\alpha \in R$⟩

and

⟨θ quasiconcave on Γ⟩ ⟺ ⟨Ω_α is convex for each $\alpha \in R$⟩

PROOF   We shall establish the theorem for the quasiconvex case.   The quasiconcave case follows from it.

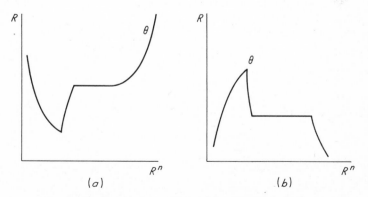

Fig. 9.1.1 A quasiconvex and a quasiconcave function on R.   (*a*) Quasiconvex θ on $R^n = R$; (*b*) quasiconcave θ on $R^n = R$.

($\Leftarrow$)   Let $x^1, x^2 \in \Gamma$, $\theta(x^2) \leq \theta(x^1)$, and $0 \leq \lambda \leq 1$. If we let $\alpha = \theta(x^1)$, then by the convexity of Λ_α we have that

$$\theta[(1 - \lambda)x^1 + \lambda x^2] \leq \alpha = \theta(x^1)$$

and hence θ is quasiconvex on Γ.

($\Rightarrow$)   Let θ be quasiconvex on Γ, α be any real number, and let x^1 and x^2 be any two points in Λ_α (if Λ_α is empty, then it is convex). Without loss of generality let $\theta(x^2) \leq \theta(x^1)$. Since $x^1, x^2 \in \Lambda_\alpha$, we have that

$$\theta(x^2) \leq \theta(x^1) \leq \alpha$$

Since θ is quasiconvex, and since Γ is convex, we have that for $0 \leq \lambda \leq 1$

$$\theta[(1 - \lambda)x^1 + \lambda x^2] \leq \theta(x^1) \leq \alpha$$

Hence $(1 - \lambda)x^1 + \lambda x^2 \in \Lambda_\alpha$, and Λ_α is convex.   ∎

4   **Theorem (differentiable quasiconvex and quasiconcave functions)**

Let Γ be an open set in R^n, and let θ be a numerical function defined on Γ. Then

$$\left\langle \begin{matrix} x^1, x^2 \in \Gamma \\ \theta \text{ differentiable and} \\ \text{quasiconvex at } x^1 \end{matrix} \right\rangle \Rightarrow \langle \theta(x^2) \leq \theta(x^1) \Rightarrow \nabla\theta(x^1)(x^2 - x^1) \leq 0 \rangle$$

$$\langle \theta \text{ quasiconvex on } \Gamma \rangle \Leftarrow \left\langle \begin{matrix} \Gamma \text{ convex, } \theta \text{ differentiable on } \Gamma \\ \left\langle \begin{matrix} x^1, x^2 \in \Gamma \\ \theta(x^2) \leq \theta(x^1) \end{matrix} \right\rangle \Rightarrow \nabla\theta(x^1)(x^2 - x^1) \leq 0 \end{matrix} \right\rangle$$

$$\left\langle \begin{matrix} x^1, x^2 \in \Gamma \\ \theta \text{ differentiable and} \\ \text{quasiconcave at } x^1 \end{matrix} \right\rangle \Rightarrow \langle \theta(x^2) \geq \theta(x^1) \Rightarrow \nabla\theta(x^1)(x^2 - x^1) \geq 0 \rangle$$

$$\langle \theta \text{ quasiconcave on } \Gamma \rangle \Leftarrow \left\langle \begin{matrix} \Gamma \text{ convex, } \theta \text{ differentiable on } \Gamma \\ \left\langle \begin{matrix} x^1, x^2 \in \Gamma \\ \theta(x^2) \geq \theta(x^1) \end{matrix} \right\rangle \Rightarrow \nabla\theta(x^1)(x^2 - x^1) \geq 0 \end{matrix} \right\rangle$$

PROOF   We shall prove the quasiconvex case only. The other case is similar.

($\Rightarrow$)   If $x^1 = x^2$, the implication is trivial. Assume then that $x^1 \neq x^2$. Since Γ is open, there exists an open ball $B_\delta(x^1)$ around x^1 which is contained in Γ. Then, for some $\bar{\mu}$ such that $0 < \bar{\mu} < 1$ and $\bar{\mu} < \delta/$

$\|x^2 - x^1\|$, we have that

$\tilde{x} = x^1 + \tilde{\mu}(x^2 - x^1) = (1 - \tilde{\mu})x^1 + \tilde{\mu}x^2 \in B_\delta(x^1) \subset \Gamma$

Then

$$\theta(x^2) \leqq \theta(x^1) \Rightarrow \theta(\tilde{x}) \leqq \theta(x^1) \qquad \text{(by 1)}$$

$$\Rightarrow \left\langle \begin{array}{c} \theta[(1 - \lambda)x^1 + \lambda\tilde{x}] \leqq \theta(x^1) \\ \text{for } 0 < \lambda \leqq 1 \end{array} \right\rangle$$

$$\text{(by 1 and convexity of } B_\delta(x^1))$$

$$\Rightarrow \left\langle \begin{array}{c} \lambda\nabla\theta(x^1)(\tilde{x} - x^1) + \alpha[x^1, \lambda(\tilde{x} - x^1)]\lambda\|\tilde{x} - x^1\| \leqq 0 \\ \text{for } 0 < \lambda \leqq 1 \\ \text{where } \lim_{\lambda \to 0} \alpha[x^1, \lambda(x^2 - x^1)] = 0 \end{array} \right\rangle$$

$$\text{(by differentiability of } \theta \text{ at } x^1)$$

$$\Rightarrow \left\langle \begin{array}{c} \nabla\theta(x^1)(\tilde{x} - x^1) + \alpha[x^1, \lambda(\tilde{x} - x^1)]\|\tilde{x} - x^1\| \leqq 0 \\ \text{for } 0 < \lambda \leqq 1 \end{array} \right\rangle$$

$$\Rightarrow \nabla\theta(x^1)(\tilde{x} - x^1) \leqq 0 \qquad \text{(by letting } \lambda \to 0)$$

$$\Rightarrow \nabla\theta(x^1)(x^2 - x^1) \leqq 0 \qquad \text{(since } \tilde{\mu} > 0)$$

$(\Leftarrow)$ Let $x^1, x^2 \in \Gamma$, let $\theta(x^2) \leqq \theta(x^1)$, let

$(x^1, x^2) = \{x \mid x = (1 - \lambda)x^1 + \lambda x^2, 0 < \lambda < 1\}$

and let

$\Omega = \{x \mid \theta(x^1) < \theta(x), x \in (x^1, x^2)\}$

We establish now the quasiconvexity of θ on Γ by showing that Ω is empty. We assume that there is an $\tilde{x} \in \Omega$ and show that a contradiction ensues. Since $\theta(x^2) \leqq \theta(x^1) < \theta(x)$ for $x \in \Omega$, we have from the hypothesis of the theorem that

$\nabla\theta(x)(x^1 - x) \leqq 0 \qquad \text{for } x \in \Omega$

and

$\nabla\theta(x)(x^2 - x) \leqq 0 \qquad \text{for } x \in \Omega$

or (since $x = (1 - \lambda)x^1 + \lambda x^2, 0 < \lambda < 1$ for $x \in \Omega$)

$$\left. \begin{array}{c} -\lambda\nabla\theta(x)(x^2 - x^1) \leqq 0 \\ (1 - \lambda)\nabla\theta(x)(x^2 - x^1) \leqq 0 \end{array} \right\rangle \text{ for } x \in \Omega$$

Since $0 < \lambda < 1$, it follows then that

$\nabla\theta(x)(x^2 - x^1) = 0 \qquad \text{for } x \in \Omega$

Since $\theta(x^1) < \theta(\bar{x})$, and since θ is continuous on Γ, the set Ω is open relative to (x^1, x^2), it contains $\bar{x}$, and there exists an $x^3 = (1 - \mu)\bar{x} + \mu x^1$, $0 < \mu \leqq 1$, such that x^3 is a point of closure of Ω, and such that $\theta(x^3) = \theta(x^1)$ (see *C.1.1* and *B.1.3*). By the mean-value theorem *D.2.1* we have that for some $\hat{x} \in \Omega$

$$0 < \theta(\bar{x}) - \theta(x^1) = \theta(\bar{x}) - \theta(x^3) = \nabla\theta(\hat{x})(\bar{x} - x^3) = \mu\nabla\theta(\hat{x})(\bar{x} - x^1)$$

and since

$$\bar{x} = (1 - \bar{\lambda})x^1 + \bar{\lambda}x^2 \qquad \text{for some } \bar{\lambda}, \, 0 < \bar{\lambda} < 1$$

then

$$0 < \mu\nabla\theta(\hat{x})(\bar{x} - x^1) = \mu\bar{\lambda}\nabla\theta(\hat{x})(x^2 - x^1) \qquad \text{for some } \bar{\lambda} > 0, \, \mu > 0$$

Since $\hat{x} \in \Omega$, the last relation above contradicts the equality established earlier, $\nabla\theta(x)(x^2 - x^1) = 0$ for all $x \in \Omega$. ∎

2. Strictly quasiconvex and strictly quasiconcave functions

We begin this section by recalling the concepts of a local minimum and maximum (see *5.1.2* or *7.1.2*). Let θ be a numerical function defined on the set $\Gamma \subset R^n$. A point $\bar{x} \in \Gamma$ is said to be a *local minimum* (*maximum*) of θ if there exists an open ball $B_\delta(\bar{x})$ around $\bar{x}$ with radius $\delta > 0$ such that

$$x \in B_\delta(\bar{x}) \cap \Gamma \Rightarrow \theta(\bar{x}) \leqq \theta(x) \qquad \text{(local minimum)}$$

$$x \in B_\delta(\bar{x}) \cap \Gamma \Rightarrow \theta(x) \leqq \theta(\bar{x}) \qquad \text{(local maximum)}$$

We remark that for a numerical function θ which is quasiconvex (quasiconcave) on a convex set Γ, a local minimum (maximum) need not be a (global) minimum (maximum) over Γ. This fact is easily demonstrated by the numerical function θ defined on R as follows:

$$\theta(x) = \begin{cases} x & \text{for } x \leqq 0 \\ 0 & \text{for } 0 < x < 1 \\ x - 1 & \text{for } x \geqq 1 \end{cases}$$

It is easy to see that θ is both quasiconvex and quasiconcave on R by verifying definitions *9.1.1* and *9.1.2*. It is also easy to see that $x = \frac{1}{2}$ is both a local minimum and a local maximum, but certainly not a global minimum or maximum over R.

We introduce now a type of function which is essentially a restriction of quasiconvex (quasiconcave) functions to functions with the property that a local minimum (maximum) is also a global minimum (maximum). Such functions have been independently introduced in [Hanson 64, Martos 65, Karamardian 67]. (These functions were called *functionally convex* by Hanson and *explicitly quasiconvex* by Martos.)

1   ### Strictly quasiconvex function

A numerical function θ defined on a set $\Gamma \subset R^n$ is said to be *strictly quasiconvex at* $\bar{x} \in \Gamma$ (with respect to Γ) if for each $x \in \Gamma$ such that $\theta(x) < \theta(\bar{x})$ the function θ assumes a lower value than $\theta(\bar{x})$ on each point in the intersection of the open line segment $(\bar{x}, x)$ and Γ, or equivalently

$$\left. \begin{array}{c} x \in \Gamma \\ \theta(x) < \theta(\bar{x}) \\ 0 < \lambda < 1 \\ (1 - \lambda)\bar{x} + \lambda x \in \Gamma \end{array} \right\} \Rightarrow \theta[(1 - \lambda)\bar{x} + \lambda x] < \theta(\bar{x})$$

θ is said to be *strictly quasiconvex on* Γ if it is strictly quasiconvex at each $x \in \Gamma$.

It follows from the above definition that a numerical function θ defined on a convex set Γ is strictly quasiconvex on Γ if and only if

$$\left. \begin{array}{c} x^1, x^2 \in \Gamma \\ \theta(x^2) < \theta(x^1) \\ 0 < \lambda < 1 \end{array} \right\} \Rightarrow \theta[(1 - \lambda)x^1 + \lambda x^2] < \theta(x^1)$$

2   ### Strictly quasiconcave function

A numerical function θ defined on a set $\Gamma \subset R^n$ is said to be *strictly quasiconcave at* $\bar{x} \in \Gamma$ (with respect to Γ) if for each $x \in \Gamma$ such that $\theta(\bar{x}) < \theta(x)$ the function θ assumes a higher value than $\theta(\bar{x})$ on each point in the intersection of the open line segment $(\bar{x}, x)$ and Γ, or equivalently

$$\left. \begin{array}{c} x \in \Gamma \\ \theta(\bar{x}) < \theta(x) \\ 0 < \lambda < 1 \\ (1 - \lambda)\bar{x} + \lambda x \in \Gamma \end{array} \right\} \Rightarrow \theta(\bar{x}) < \theta[(1 - \lambda)\bar{x} + \lambda x]$$

θ is said to be *strictly quasiconcave on* Γ if it is strictly quasiconcave at each $x \in \Gamma$.

It follows from the above definition that a numerical function θ defined on a convex set Γ is strictly quasiconcave on Γ if and only if

$$\left.\begin{array}{c} x^1, x^2 \in \Gamma \\ \theta(x^1) < \theta(x^2) \\ 0 < \lambda < 1 \end{array}\right\} \Rightarrow \theta(x^1) < \theta[(1-\lambda)x^1 + \lambda x^2]$$

Obviously θ is strictly quasiconcave at $\bar{x}$ (on Γ) if and only if $-\theta$ is strictly quasiconvex at $\bar{x}$ (on Γ).

Figure *9.2.1* depicts a strictly quasiconvex and a strictly quasiconcave function on R. The quasiconvex function of Fig. *9.1.1a* is not strictly quasiconvex, and the quasiconcave function of Fig. *9.1.1b* is not strictly quasiconcave. (Why?)

We observe that a strictly quasiconvex function need not be quasiconvex. Consider the numerical function θ defined on R as follows

$$\theta(x) = \left\{\begin{array}{ll} 1 & \text{for } x = 0 \\ 0 & \text{for } x \neq 0 \end{array}\right.$$

This function is strictly quasiconvex on R but is not quasiconvex on R. For by taking $x^1 = -1$, $x^2 = 1$, $\lambda = \frac{1}{2}$, we see that $\theta(x^2) = \theta(x^1)$, but $\theta[(1-\lambda)x^1 + \lambda x^2] > \theta(x^1)$.

If we require that θ be lower semicontinuous, the above counter-example will be eliminated and every strictly quasiconvex function will also be quasiconvex. In fact we have the following result.

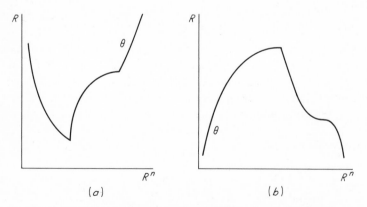

Fig. 9.2.1 Strictly quasiconvex and strictly quasiconcave functions. (a) Strictly quasiconvex function θ on $R^n = R$; (b) strictly quasiconcave function θ on $R^n = R$.

3 ### Theorem [Karamardian 67]

Let θ be a lower (upper) semicontinuous numerical function defined on the convex set Γ in R^n. If θ is strictly quasiconvex (strictly quasiconcave) on Γ, then θ is quasiconvex (quasiconcave) on Γ, but not conversely.

PROOF We prove only the quasiconvex case. Let θ be strictly quasiconvex on Γ, and let x^1 and x^2 be in Γ. By *1* we have that

$$\left.\begin{array}{c} \theta(x^2) < \theta(x^1) \\ 0 < \lambda < 1 \end{array}\right\} \Rightarrow \theta[(1 - \lambda)x^1 + \lambda x^2] < \theta(x^1)$$

Hence if $\theta(x^2) < \theta(x^1)$, we are done by *9.1.1*. Assume now that $\theta(x^2) = \theta(x^1)$. We will show (by contradiction) that there exists no $\hat{x} \in (x^1, x^2)$ such that $\theta(x^1) < \theta(\hat{x})$. This will then establish the quasiconvexity of θ by *9.1.1*. Let $\hat{x} \in (x^1, x^2)$ such that $\theta(x^1) < \theta(\hat{x})$. Then

$$\hat{x} \in \Omega = \{x \mid \theta(x^1) < \theta(x), \, x \in (x^1, x^2)\}$$

Since θ is lower semicontinuous on Γ, Ω is open relative to (x^1, x^2) by *C.1.2*(iv). Hence, there exists an $\bar{x} \in (x^1, \hat{x}) \cap \Omega$. By the strict quasiconvexity of θ and *1* we have that (since $\hat{x} \in \Omega$ and $\bar{x} \in \Omega$)

$$\theta(x^1) < \theta(\bar{x}) \Rightarrow \theta(\bar{x}) < \theta(\hat{x})$$

and

$$\theta(x^2) < \theta(\bar{x}) \Rightarrow \theta(\hat{x}) < \theta(\bar{x})$$

which is a contradiction. Hence no such $\hat{x}$ exists and θ is quasiconvex on Γ.

That the converse is not true follows from the example given at the beginning of this section, which is quasiconvex on R but not strictly quasiconvex on R. (For take $x^1 = \frac{1}{2}$, $x^2 = -\frac{1}{2}$, $\lambda = \frac{1}{10}$, then $\theta(x^2) < \theta(x^1)$, but $\theta[(1 - \lambda)x^1 + \lambda x^2] = \theta(x^1)$, which contradicts *1*.) ∎

4 ### Theorem

Let θ be a numerical function defined on the convex set Γ in R^n, and let $\bar{x} \in \Gamma$ be a local minimum (maximum). If θ is strictly quasiconvex (strictly quasiconcave) at $\bar{x}$, then $\theta(\bar{x})$ is a global minimum (maximum) of θ on Γ.

PROOF We prove the strictly quasiconvex case. If $\bar{x}$ is a local minimum, then there exists a ball $B_\delta(\bar{x})$ such that

$$x \in B_\delta(\bar{x}) \cap \Gamma \Rightarrow \theta(\bar{x}) \leq \theta(x)$$

Assume now that there exists an $\hat{x}$ in Γ but not in $B_\delta(\bar{x})$ such that $\theta(\hat{x}) < \theta(\bar{x})$. By the strict quasiconvexity of θ at $\bar{x}$ and the convexity of Γ, we have that

$$\theta[(1 - \lambda)\bar{x} + \lambda\hat{x}] < \theta(\bar{x}) \qquad \text{for } 0 < \lambda < 1$$

But for $\lambda < \delta/\|\hat{x} - \bar{x}\|$, we have that

$$(1 - \lambda)\bar{x} + \lambda\hat{x} \in B_\delta(\bar{x}) \cap \Gamma$$

and hence

$$\theta(\bar{x}) \leqq \theta[(1 - \lambda)\bar{x} + \lambda\hat{x}] \qquad \text{for } 0 < \lambda < \frac{\delta}{\|\hat{x} - \bar{x}\|}$$

which contradicts a previous inequality. ∎

In closing this section on strictly quasiconvex (strictly quasiconcave) functions, we remark that there does not seem to be a simple characterization of a differentiable strictly quasiconvex (strictly quasiconcave) function in terms of the gradient of the function [such as *9.1.4* for a quasiconvex (quasiconcave) function].

3.   Pseudoconvex and pseudoconcave functions

Let θ be a differentiable function on R^n. If $\nabla\theta(\bar{x}) = 0$, then it follows from *6.1.1* that $\theta(x) \geqq \theta(\bar{x})$ for all x in R^n if θ is convex at $\bar{x}$. If θ is a differentiable strictly quasiconvex function (and hence by *9.2.3* also a quasiconvex function) on R^n, and if $\nabla\theta(\bar{x}) = 0$, it does not necessarily follow that $\theta(x) \geqq \theta(\bar{x})$ for all x in R^n. For example take the strictly quasiconvex function θ defined on R by $\theta(x) = (x)^3$. For this function $\nabla\theta(0) = 0$, but certainly $\theta(0)$ is not a minimum of θ on R. In order to exclude such strictly quasiconvex functions that have inflection points with horizontal tangents, we introduce the concept of a pseudoconvex function. (The concept of a pseudoconvex function was introduced independently in [Mangasarian 65] and in a slightly different form under the name of *semiconvex function* in [Tuy 64]. Most of the results for pseudoconvex functions derived in this book are based on [Mangasarian 65].)

1   **Pseudoconvex function**

Let θ be a numerical function defined on some open set in R^n containing the set Γ. θ is said to be *pseudoconvex at* $\bar{x} \in \Gamma$ (with respect to Γ) if it is differentiable at $\bar{x}$ and

$$\left.\begin{array}{r} x \in \Gamma \\ \nabla\theta(\bar{x})(x - \bar{x}) \geq 0 \end{array}\right\} \Rightarrow \theta(x) \geq \theta(\bar{x})$$

θ is said to be *pseudoconvex on* Γ if it is pseudoconvex at each $x \in \Gamma$.

2   **Pseudoconcave function**

Let θ be a numerical function defined on some open set in R^n containing the set Γ. θ is said to be *pseudoconcave at* $\bar{x} \in \Gamma$ (with respect to Γ) if it is differentiable at $\bar{x}$ and

$$\left.\begin{array}{r} x \in \Gamma \\ \nabla\theta(\bar{x})(x - \bar{x}) \leq 0 \end{array}\right\} \Rightarrow \theta(x) \leq \theta(\bar{x})$$

θ is said to be *pseudoconcave on* Γ if it is pseudoconcave at each $x \in \Gamma$.

Obviously θ is pseudoconcave at $\bar{x}$ (on Γ) if and only if $-\theta$ is pseudoconvex at $\bar{x}$ (on Γ).

Figure *9.3.1* depicts a pseudoconvex function and a pseudoconcave function on R.

We give now some results involving pseudoconvex and pseudoconcave functions.

3   **Theorem**

Let θ be a numerical function defined on some open set in R^n containing the set Γ, and let $\bar{x} \in \Gamma$.

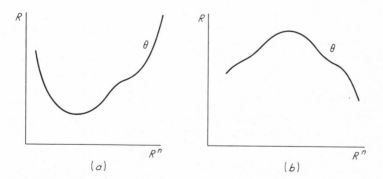

Fig. 9.3.1 Pseudoconvex and pseudoconcave functions. (*a*) Pseudoconvex function θ on $R^n = R$; (*b*) pseudoconcave function θ on $R^n = R$.

(i) *Let Γ be convex, and let θ be differentiable at $\bar{x}$. Then*

$$\theta(\bar{x}) = \min_{x \in \Gamma} \theta(x) \Rightarrow \nabla\theta(\bar{x})(x - \bar{x}) \geq 0 \qquad \text{for all } x \in \Gamma$$

$$[\theta(\bar{x}) = \max_{x \in \Gamma} \theta(x) \Rightarrow \nabla\theta(\bar{x})(x - \bar{x}) \leq 0 \qquad \text{for all } x \in \Gamma]$$

(ii) *Let θ be pseudoconvex (pseudoconcave) at $\bar{x}$. Then*

$$\theta(\bar{x}) = \min_{x \in \Gamma} \theta(x) \Leftarrow \nabla\theta(\bar{x})(x - \bar{x}) \geq 0 \qquad \text{for all } x \in \Gamma$$

$$[\theta(\bar{x}) = \max_{x \in \Gamma} \theta(x) \Leftarrow \nabla\theta(\bar{x})(x - \bar{x}) \leq 0 \qquad \text{for all } x \in \Gamma]$$

PROOF   (i)   Let x be any point in Γ.   Since Γ is convex

$$(1 - \lambda)\bar{x} + \lambda x \in \Gamma \qquad \text{for } 0 \leq \lambda \leq 1$$

Since θ is differentiable at $\bar{x}$, and since $\theta(\bar{x}) = \min_{x \in \Gamma} \theta(x)$, we have that

$$0 \leq \theta[(1 - \lambda)\bar{x} + \lambda x] - \theta(\bar{x}) = \lambda\nabla\theta(\bar{x})(x - \bar{x}) + \alpha[\bar{x}, \lambda(x - \bar{x})]\lambda\|x - \bar{x}\|$$

where

$$\lim_{\lambda \to 0} \alpha[\bar{x}, \lambda(x - \bar{x})] = 0$$

Hence

$$\nabla\theta(\bar{x})(x - \bar{x}) + \alpha[\bar{x}, \lambda(x - \bar{x})]\|x - \bar{x}\| \geq 0 \qquad \text{for } 0 < \lambda \leq 1$$

By taking the limit as λ approaches zero, we get

$$\nabla\theta(\bar{x})(x - \bar{x}) \geq 0$$

The second implication of (i) follows similarly.

(ii)   For any x in Γ we have that $\nabla\theta(\bar{x})(x - \bar{x}) \geq 0$, hence by *1*, $\theta(x) \geq \theta(\bar{x})$ and $\theta(\bar{x}) = \min_{x \in \Gamma} \theta(x)$.   The second implication follows from *2*. ∎

4   **Corollary**

Let θ be a numerical function defined on the open convex set Γ in R^n. Let $\bar{x} \in \Gamma$, and let θ be differentiable at $\bar{x}$.   Then

(i) $\theta(\bar{x}) = \min_{x \in \Gamma} \theta(x) \Rightarrow \nabla\theta(\bar{x}) = 0$

$[\theta(\bar{x}) = \max_{x \in \Gamma} \theta(x) \Rightarrow \nabla\theta(\bar{x}) = 0]$

(ii)   *Let θ be pseudoconvex (pseudoconcave) at $\bar{x}$.   Then*

$$\theta(\bar{x}) = \min_{x \in \Gamma} \theta(x) \Leftarrow \nabla\theta(\bar{x}) = 0$$

$$[\theta(\bar{x}) = \max_{x \in \Gamma} \theta(x) \Leftarrow \nabla\theta(\bar{x}) = 0]$$

PROOF   (i)   We consider only the first implication.   By 3(i) we have that $\nabla\theta(\bar{x})(x - \bar{x}) \geqq 0$ for all $x \in \Gamma$.   Since Γ is open, $x = \bar{x} - \delta\nabla\theta(\bar{x}) \in \Gamma$ for some $\delta > 0$.   Hence $-\delta\nabla\theta(\bar{x})\nabla\theta(\bar{x}) \geqq 0$, which implies that $\nabla\theta(\bar{x}) = 0$.

(ii)   Follows trivially from 3(ii).   ∎

We relate now pseudoconvex functions to strictly quasiconvex functions, quasiconvex functions, and convex functions.

5   **Theorem**

Let Γ be a convex set in R^n, and let θ be a numerical function defined on some open set containing Γ.   If θ is pseudoconvex (pseudoconcave) on Γ, then θ is strictly quasiconvex (strictly quasiconcave) on Γ and hence also quasiconvex (quasiconcave) on Γ.   The converse is not true.

PROOF   Let θ be pseudoconvex on Γ.   We shall assume that θ is not strictly quasiconvex on Γ and exhibit a contradiction.   If θ is not strictly quasiconvex on Γ, then it follows from $9.2.1$ that there exist $x^1, x^2 \in \Gamma$ such

$$\theta(x^2) < \theta(x^1)$$

and

$$\theta(\hat{x}) \geqq \theta(x^1)$$

for some $\hat{x}$ such that

$$\hat{x} \in (x^1, x^2) = \{x \mid x = (1 - \lambda)x^1 + \lambda x^2, 0 < \lambda < 1\}$$

Hence there exists an $\bar{x} \in (x^1, x^2)$ such that

$$\theta(\bar{x}) = \max_{x \in [x^1, x^2]} \theta(x)$$

where

$$[x^1, x^2] = \{x \mid x = (1 - \lambda)x^1 + \lambda x^2, 0 \leq \lambda \leq 1\}$$

By Theorem 3(i) above we have then that

$$\nabla\theta(\bar{x})(x^1 - \bar{x}) \leqq 0$$

and

$$\nabla\theta(\bar{x})(x^2 - \bar{x}) \leqq 0$$

Since

$$\bar{x} = (1 - \bar{\lambda})x^1 + \bar{\lambda}x^2 \qquad \text{for some } \bar{\lambda}, \, 0 < \bar{\lambda} < 1$$

then

$$0 \geqq \nabla\theta(\bar{x})(x^1 - \bar{x}) = \bar{\lambda}\nabla\theta(\bar{x})(x^1 - x^2)$$

and

$$0 \geqq \nabla\theta(\bar{x})(x^2 - \bar{x}) = -(1 - \bar{\lambda})\nabla\theta(\bar{x})(x^1 - x^2)$$

Hence

$$\nabla\theta(\bar{x})(x^1 - x^2) = 0$$

and

$$\nabla\theta(\bar{x})(x^2 - \bar{x}) = -(1 - \bar{\lambda})\nabla\theta(\bar{x})(x^1 - x^2) = 0$$

But by the pseudoconvexity of θ on Γ, it follows that

$$\theta(x^2) \geqq \theta(\bar{x}) \qquad (\text{since } \nabla\theta(\bar{x})(x^2 - \bar{x}) = 0)$$

and hence

$$\theta(x^1) > \theta(\bar{x}) \qquad (\text{since } \theta(x^1) > \theta(x^2))$$

This last inequality contradicts the earlier statement that

$$\theta(\bar{x}) = \max_{x \in [x^1, x^2]} \theta(x)$$

Hence θ is strictly quasiconvex on Γ and, by *9.2.3*, is also quasiconvex on Γ. That the converse is not necessarily true can be seen from the example $\theta(x) = (x)^3$, $x \in R$, which is strictly quasiconvex on R but is not pseudoconvex on R. ∎

6   **Theorem**

Let θ be a numerical function defined on some open Γ set in R^n, let $\bar{x} \in \Gamma$, and let θ be differentiable at $\bar{x}$. If θ is convex (concave) at $\bar{x}$, then θ is pseudoconvex (pseudoconcave) at $\bar{x}$, but not conversely.

PROOF Let θ be convex at $\bar{x}$. By *6.1.1* we have that

$$\theta(x) - \theta(\bar{x}) \geqq \nabla\theta(\bar{x})(x - \bar{x}) \qquad \text{for each } x \in \Gamma$$

Hence

$$\left.\begin{array}{c} x \in \Gamma \\ \nabla \theta(\bar{x})(x - \bar{x}) \geqq 0 \end{array}\right\} \Rightarrow \theta(x) \geqq \theta(\bar{x})$$

and θ is pseudoconvex at $\bar{x}$. That the converse is not necessarily true can be seen from the example

$$\theta(x) = x + (x)^3 \qquad x \in R$$

θ is not convex on R because $\nabla^2 \theta(x) < 0$ for $x < 0$. However, θ is pseudoconvex on R because $\nabla \theta(x) = 1 + 3(x)^2 > 0$ and

$$\nabla \theta(\bar{x})(x - \bar{x}) \geqq 0 \Rightarrow x - \bar{x} \geqq 0$$

$$\Rightarrow (x)^3 \geqq (\bar{x})^3$$

$$\Rightarrow \theta(x) - \theta(\bar{x}) = x + (x)^3 - \bar{x} - (\bar{x})^3 \geqq 0 \quad \blacksquare$$

7   Theorem

Let Γ be a convex set in R^n, and let θ be a numerical function defined on some open set containing Γ. If θ is pseudoconvex (pseudoconcave) on Γ, then each local minimum (maximum) of θ on Γ is also a global minimum (maximum) of θ on Γ.

PROOF   By Theorem 5 θ is strictly quasiconvex (strictly quasiconcave) on Γ. By Theorem 9.2.4 each local minimum (maximum) of θ on Γ is also a global minimum (maximum) on Γ. $\quad \blacksquare$

4. Summary of properties and relations between quasiconvex, strictly quasiconvex, pseudoconvex, convex, and strictly convex functions

In Fig. 9.4.1 we summarize the relations between differentiable functions which are quasiconvex, strictly quasiconvex, pseudoconvex, convex, and strictly convex, all defined on an open convex set Γ in R^n. For convenience, we include statements which are logical equivalents of each other. For example, we include the two following equivalent definitions of a pseudoconvex function

$$\theta(x^2) < \theta(x^1) \Rightarrow \nabla \theta(x^1)(x^2 - x^1) < 0$$

or

$$\theta(x^2) \geqq \theta(x^1) \Leftarrow \nabla \theta(x^1)(x^2 - x^1) \geqq 0$$

In Fig. $9.4.1$ certain inequalities hold only if $x^1 \neq x^2$ and $\lambda \neq 0$ or $\lambda \neq 1$. All implications are represented by single arrows. Implications not so represented are not true in general. Figure $9.4.1$ immediately shows that the class of quasiconvex functions is the largest class considered and the strictly convex class is the smallest. Figure $9.4.2$ shows similar results for differentiable quasiconcave functions, etc.

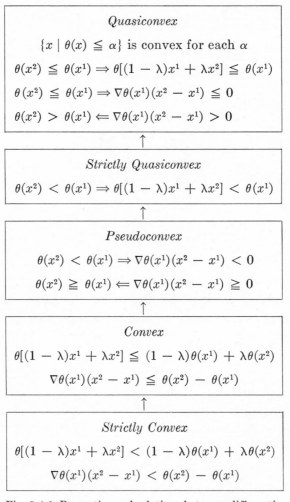

Fig. 9.4.1 Properties and relations between differentiable quasiconvex, strictly quasiconvex, pseudoconvex, convex and strictly convex functions defined on an open convex set $\Gamma \subset R^n$.

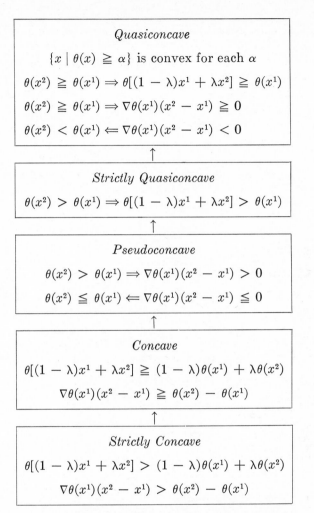

Fig. 9.4.2 Properties and relations between differentiable quasiconcave, strictly quasiconcave, pseudoconcave, concave and strictly concave functions defined on an open convex set $\Gamma \subset R^n$.

5. Warning

It was shown in *4.1.6* that each nonnegative linear combination of convex functions is also convex. This is not true in general for quasiconvex, strictly quasiconvex, or pseudoconvex functions. We indicate the reason for the pseudoconvex functions θ and ϕ on R^n. If we define

$$\psi(x) = \theta(x) + \phi(x)$$

then the requirement that

$$\nabla\psi(x^1)(x^2 - x^1) \geqq 0$$

does not ensure that

$$\nabla\theta(x^1)(x^2 - x^1) \geqq 0 \qquad \text{and} \qquad \nabla\phi(x^1)(x^2 - x^1) \geqq 0$$

and hence we cannot conclude that

$$\theta(x^2) \geqq \theta(x^1), \; \phi(x^2) \geqq \phi(x^1), \; \text{and hence} \; \psi(x^2) \geqq \psi(x^1)$$

and so ψ may not be pseudoconvex on R^n. For example $\theta(x) = -x$ and $\phi(x) = x + (x)^3$ are both pseudoconvex on R, but their sum $\psi(x) = (x)^3$ is not pseudoconvex on R.

6. Problems

1   Nonlinear fractional functions

Let ϕ and ψ be numerical functions defined on a set $\Gamma \subset R^n$, and let $\psi \neq 0$ on Γ.

(i)   Let Γ be open, let $\bar{x} \in \Gamma$, and let ϕ and ψ be differentiable at $\bar{x}$. Show that $\theta = \phi/\psi$ is pseudoconvex at $\bar{x}$ if

$$\left\langle \begin{array}{l} \phi \text{ is convex at } \bar{x}, \psi > 0 \text{ on } \Gamma, \text{ or} \\ \phi \text{ is concave at } \bar{x}, \psi < 0 \text{ on } \Gamma \end{array} \right\rangle$$

and

$$\left\langle \begin{array}{l} \psi \text{ is linear on } R^n, \text{ or} \\ \psi \text{ is convex at } \bar{x}, \phi \leqq 0 \text{ on } \Gamma, \text{ or} \\ \psi \text{ is concave at } \bar{x}, \phi \geqq 0 \text{ on } \Gamma \end{array} \right\rangle$$

(Hint: Use *6.1.1* and *9.3.1*.)

(ii)   Let Γ be convex. Show that $\theta = \phi/\psi$ is quasiconvex on Γ if

$$\left\langle \begin{array}{l} \phi \text{ is convex on } \Gamma, \psi > 0 \text{ on } \Gamma, \text{ or} \\ \phi \text{ is concave on } \Gamma, \psi < 0 \text{ on } \Gamma \end{array} \right\rangle$$

and

$$\left\langle \begin{array}{l} \psi \text{ is linear on } R^n, \text{ or} \\ \psi \text{ is convex on } \Gamma, \phi \leqq 0 \text{ on } \Gamma, \text{ or} \\ \psi \text{ is concave on } \Gamma, \phi \geqq 0 \text{ on } \Gamma \end{array} \right\rangle$$

(Hint: Consider, for the case $\psi > 0$, the set

$$\Lambda_\alpha = \{x \mid x \in \Gamma, \theta(x) \leqq \alpha\} = \{x \mid x \in \Gamma, \phi(x) - \alpha\psi(x) \leqq 0\}$$

for any real α, and use *9.1.3* and *4.1.6*.)

2   Nonlinear fractional functions

Show that the words convex and concave can be interchanged throughout the above problem *1*, except that the sentence "Let Γ be convex . . ." remains unchanged.

3   Linear fractional functions

Let $a \in R^n$, $b \in R^n$, $\alpha \in R$, $\beta \in R$. Show that the function θ defined by

$$\theta(x) = \frac{ax + \alpha}{bx + \beta}$$

is both pseudoconvex and pseudoconcave (and hence also both quasiconvex and quasiconcave) on each convex set $\Gamma \subset R^n$ on which $bx + \beta \neq 0$.

4   Bi-nonlinear functions

Let ϕ and σ be numerical functions defined on a convex set $\Gamma \in R^n$, and let $\sigma \neq 0$ on Γ.

(i)   Let Γ be open, and let ϕ and σ be differentiable on Γ. Show that $\theta = \phi\sigma$ is pseudoconvex on Γ if

$\langle\phi$ is convex, $\phi \leqq 0$, σ is concave, $\sigma > 0$, all on $\Gamma\rangle$

or

$\langle\phi$ is concave, $\phi \geqq 0$, σ is convex, $\sigma < 0$, all on $\Gamma\rangle$

Show that θ is pseudoconcave on Γ if

$\langle\phi$ is convex, $\phi \leqq 0$, σ is convex, $\sigma < 0$, all on $\Gamma\rangle$

or

$\langle\phi$ is concave, $\phi \geqq 0$, σ is concave, $\sigma > 0$, all on $\Gamma\rangle$

(ii) If we drop the assumptions that Γ is open and that ϕ and σ are differentiable on Γ, show that either of the first two conditions of (i) make θ quasiconvex on Γ and either of the last two conditions of (i) make θ quasiconcave on Γ. (Hint: Show first that the reciprocal of a positive concave function is a positive convex function, and that the reciprocal of a negative convex function is a negative concave function, then use the results of Prob. *1* above.)

5   Let θ be a positive numerical function defined on the open convex set Γ in R^n. Show that if $\log \theta$ is convex (concave) on Γ, then θ is convex (pseudoconcave) on Γ.

Chapter Ten

Optimality and Duality for Generalized Convex and Concave Functions

In this chapter we shall use the concepts of quasiconvexity (quasiconcavity) and pseudoconvexity (pseudoconcavity), introduced in the previous chapter, to derive sufficient optimality criteria, necessary optimality criteria, and duality relations for a class of nonlinear programming problems that involve differentiable functions. The pioneering work in this direction was that of [Arrow-Enthoven 61]. Subsequent work appeared in [Tuy 64, Martos 65, Mangasarian 65].

1. Sufficient optimality criteria

We establish in this section two sufficient optimality criteria which are slight generalizations of the sufficiency result of [Mangasarian 65]. The present results also subsume the sufficiency results of [Arrow-Enthoven 61] and Theorem *7.2.1* of this book.

1 **Sufficient optimality theorem**

Let X^0 be an open set in R^n, and let θ and g be respectively a numerical function and an m-dimensional vector function both defined on X^0. Let $\bar{x} \in X^0$, let

$$I = \{i \mid g_i(\bar{x}) = 0\}$$

let θ be pseudoconvex at $\bar{x}$, and let g_I be differentiable and quasiconvex at $\bar{x}$. If there exists a $\bar{u} \in R^m$ such that $(\bar{x},\bar{u})$ satisfies the following conditions

$$[\nabla\theta(\bar{x}) + \bar{u}\nabla g(\bar{x})](x - \bar{x}) \geq 0 \qquad \text{for all } x \in X$$

$$\bar{u}g(\bar{x}) = 0$$

$$g(\bar{x}) \leq 0$$

$$\bar{u} \geq 0$$

then $\bar{x}$ is a solution of the following minimization problem

$$\theta(\bar{x}) = \min_{x \in X} \theta(x) \qquad \bar{x} \in X = \{x \mid x \in X^0, g(x) \leq 0\}$$

PROOF   Let

$$I = \{i \mid g_i(\bar{x}) = 0\} \qquad J = \{i \mid g_i(\bar{x}) < 0\} \qquad I \cup J = \{1, 2, \ldots, m\}$$

Since $\bar{u} \geq 0$, $g(\bar{x}) \leq 0$, and $\bar{u}g(\bar{x}) = 0$, we have that

$$\bar{u}_i g_i(\bar{x}) = 0 \qquad \text{for } i = 1, \ldots, m$$

and hence

$$\bar{u}_J = 0$$

Since

$$g_I(x) \leq 0 = g_I(\bar{x}) \qquad \text{for all } x \in X$$

it follows by the quasiconvexity of g_I at $\bar{x}$ and Theorem *9.1.4* that

$$\nabla g_I(\bar{x})(x - \bar{x}) \leq 0 \qquad \text{for all } x \in X$$

and hence

$$\bar{u}_I \nabla g_I(\bar{x})(x - \bar{x}) \leq 0 \qquad \text{for all } x \in X$$

But since $\bar{u}_J = 0$, we also have that

$$\bar{u}_J \nabla g_J(\bar{x})(x - \bar{x}) = 0 \qquad \text{for all } x \in X$$

Addition of the last two relations gives

$$\bar{u}\nabla g(\bar{x})(x - \bar{x}) = [\bar{u}_I \nabla g_I(\bar{x}) + \bar{u}_J \nabla g_J(\bar{x})](x - \bar{x}) \leq 0 \qquad \text{for all } x \in X$$

But since $[\nabla\theta(\bar{x}) + \bar{u}\nabla g(\bar{x})](x - \bar{x}) \geq 0$ for all $x \in X$, the last inequality gives

$$\nabla\theta(\bar{x})(x - \bar{x}) \geq 0 \qquad \text{for all } x \in X$$

which, by the pseudoconvexity of θ at $\bar{x}$, implies that

$$\theta(x) \geq \theta(\bar{x}) \qquad \text{for all } x \in X$$

Since $g(\bar{x}) \leq 0$, we have that $\bar{x} \in X$, and

$$\theta(\bar{x}) = \min_{x \in X} \theta(x) \qquad \bar{x} \in X \quad \blacksquare$$

The following theorem, which is a generalizaton of the Kuhn-Tucker sufficient optimality criterion 7.2.1, follows trivially from the above theorem by observing that $\nabla\theta(\bar{x}) + \bar{u}\nabla g(\bar{x}) = 0$ implies that $[\nabla\theta(\bar{x}) + \bar{u}\nabla g(\bar{x})](x - \bar{x}) = 0$ for all $x \in X$.

2   Kuhn-Tucker sufficient optimality theorem

Theorem 1 above holds with the condition

$$\nabla\theta(\bar{x}) + \bar{u}\nabla g(\bar{x}) = 0$$

replacing the condition

$$[\nabla\theta(\bar{x}) + \bar{u}\nabla g(\bar{x})](x - \bar{x}) \geq 0 \qquad \text{for all } x \in X$$

2.   Necessary optimality criteria

The purpose of this section is to generalize the necessary optimality criteria of Sec. 7.3 by employing the concepts introduced in Chap. 9. We begin by extending Abadie's linearization lemma 7.3.1.

1   Linearization lemma

Let X^0 be an open set in R^n, and let θ and g be respectively a numerical function and an m-dimensional vector function both defined on X^0. Let $\bar{x}$ be a solution of LMP 7.1.2, let θ and g be differentiable at $\bar{x}$, let

$$P = \{i \mid g_i(\bar{x}) = 0, \text{ and } g_i \text{ is pseudoconcave at } \bar{x}\}$$

and let

$$Q = \{i \mid g_i(\bar{x}) = 0, \text{ and } g_i \text{ is not pseudoconcave at } \bar{x}\}$$

Then the system

$$\left\langle \begin{array}{l} \nabla\theta(\bar{x})z < 0 \\ \nabla g_Q(\bar{x})z < 0 \\ \nabla g_P(\bar{x})z \leq 0 \end{array} \right\rangle$$

has no solution z in R^n.

PROOF   The proof is identical to that of Lemma 7.3.1 except that V and W are replaced by P and Q and step (iii) of the proof of 7.3.1 is modified as follows:

(iii′) For $i \in P$, we have by *9.3.2*, since g_i is pseudoconcave at $\bar{x}$ and $\delta \nabla g_P(\bar{x}) z \leqq 0$, that

$$g_i(\bar{x} + \delta z) \leqq g_i(\bar{x}) = 0 \qquad \text{for } 0 < \delta < \hat{\delta} \text{ and } i \in P \quad \blacksquare$$

We generalize now the Fritz John stationary-point necessary optimality theorem *7.3.2*.

2

Fritz John stationary-point necessary optimality theorem

Let $\bar{x}$ be a solution of LMP 7.1.2 or of MP 7.1.1, let X^0 be open, and let θ and g be differentiable at $\bar{x}$. Then there exist an $\bar{r}_0 \in R$ and an $\bar{r} \in R^m$ such that $(\bar{x}, \bar{r}_0, \bar{r})$ solves FJP 7.1.3 and

$$(\bar{r}_0, \bar{r}_Q) \geq 0$$

where

$$Q = \{i \mid g_i(\bar{x}) = 0, \text{ and } g_i \text{ is not pseudoconcave at } \bar{x}\}$$

PROOF The proof follows from Lemma *1* above in exactly the same way as Theorem *7.3.2* follows from Lemma *7.3.1*. $\blacksquare$

Again, as in Chap. 7, there is no guarantee that $\bar{r}_0 > 0$ in the above theorem. To ensure this, we impose constraint qualifications. We give below less restrictive versions of constraint qualifications that were introduced earlier.

3

The weak Arrow-Hurwicz-Uzawa constraint qualification
(see *7.3.4*)

Let X^0 be an open set in R^n, let g be an m-dimensional vector function defined on X^0, and let

$$X = \{x \mid x \in X^0, g(x) \leqq 0\}$$

g is said to satisfy the *weak Arrow-Hurwicz-Uzawa constraint qualification at $\bar{x} \in X$* if g is differentiable at $\bar{x}$ and

$$\left\langle \begin{array}{c} \nabla g_Q(\bar{x}) z > 0 \\ \nabla g_P(\bar{x}) z \geqq 0 \end{array} \right\rangle \text{ has a solution } z \in R^n$$

where

$$P = \{i \mid g_i(\bar{x}) = 0, \text{ and } g_i \text{ is pseudoconcave at } \bar{x}\}$$

and

$$Q = \{i \mid g_i(\bar{x}) = 0, \text{ and } g_i \text{ is not pseudoconcave at } \bar{x}\}$$

4   **The weak reverse convex constraint qualification** (see *7.3.5*)

Let X^0 be an open set in R^n, let g be an m-dimensional vector function defined on X^0, and let

$$X = \{x \mid x \in X^0, g(x) \leq 0\}$$

g is said to satisfy the *weak reverse convex constraint qualification at* $\bar{x} \in X$ if g is differentiable at $\bar{x}$ and, for each $i \in I$, either g_i is pseudo-concave at $\bar{x}$, or g_i is linear on R^n, where

$$I = \{i \mid g_i(\bar{x}) = 0\}$$

5   **Slater's weak constraint qualification** (see *5.4.3*)

Let X^0 be an open set in R^n, let g be an m-dimensional vector function defined on X^0, and let

$$X = \{x \mid x \in X^0, g(x) \leq 0\} \ \cdot$$

g is said to satisfy *Slater's weak constraint qualification at* $\bar{x} \in X$ if g is differentiable at $\bar{x}$, g_I is pseudoconvex at $\bar{x}$, and there exists an $\hat{x} \in X$ such that $g_I(\hat{x}) < 0$ where

$$I = \{i \mid g_i(\bar{x}) = 0\}$$

We observe here that each of the weak constraint qualifications *3*, *4*, and *5* above is essentially implied by the corresponding constraint qualification introduced earlier, that is, by *7.3.4*, *7.3.5*, and *5.4.3*, respectively.

Before deriving the Kuhn-Tucker necessary optimality conditions under the weakened constraint qualifications above, we relate the weakened constraint qualifications to each other and to the Kuhn-Tucker constraint qualification *7.3.3*.

6   **Lemma**

Let X^0 be an open set in R^n, let g be an m-dimensional vector function defined on X^0, and let

$$X = \{x \mid x \in X^0, g(x) \leq 0\}$$

(i)   *If g satisfies the weak reverse convex constraint qualification 4 at $\bar{x}$, then g satisfies the weak Arrow-Hurwicz-Uzawa constraint qualification 3 at $\bar{x}$.*

(ii) *If g satisfies the weak reverse convex constraint qualification 4 at $\bar{x}$, then g satisfies the Kuhn-Tucker constraint qualification 7.3.3 at $\bar{x}$.*

(iii)   *If g satisfies Slater's weak constraint qualification 5 at $\bar{x}$, then g satisfies the weak Arrow-Hurwicz-Uzawa constraint qualification 3 at $\bar{x}$.*

PROOF   (i)   We have here that Q is empty, and hence by taking $z = 0$, we immediately satisfy $\nabla g_p(\bar{x})z = 0$, and the weak Arrow-Hurwicz-Uzawa constraint qualification *3* is satisfied at $\bar{x}$.

(ii)   The proof is essentially identical to the proof of Lemma *7.3.6*(ii), except that the line reading

$$g_I[e(\tau)] = g_I(\bar{x} + \lambda\tau y) \leqq g_I(\bar{x}) + \lambda\tau\nabla g_I(\bar{x}) = \lambda\tau\nabla g_I(\bar{x})y \leqq 0$$

in the proof of *7.3.6*(ii) is replaced by the following argument:

$$\lambda\tau\nabla g_I(\bar{x})y \leqq 0 \Rightarrow g_I(\bar{x} + \lambda\tau y) - g_I(\bar{x}) \leqq 0 \qquad \text{(by pseudoconcavity of } g_I \text{ at } \bar{x})$$

$$\Rightarrow g_I(\bar{x} + \lambda\tau y) \leqq 0 \qquad \text{(since } g_I(\bar{x}) = 0)$$

$$\Rightarrow g_I[e(\tau)] \leqq 0 \qquad \text{(since } e(\tau) = \bar{x} + \lambda\tau y)$$

(iii)   Let g satisfy Slater's weak constraint qualification *5* at $\bar{x}$. Then there exists an $\hat{x} \in X$ such that

$$0 > g_I(\hat{x}) = g_I(\hat{x}) - g_I(\bar{x})$$

where

$$I = \{i \mid g_i(\bar{x}) = 0\}$$

But since g_I is pseudoconvex at $\bar{x}$, it follows from (the logical equivalent of) *9.3.1* that

$$g_I(\hat{x}) < g_I(\bar{x}) \Rightarrow \nabla g_I(\bar{x})(\hat{x} - \bar{x}) < 0$$

Hence by taking $z = \bar{x} - \hat{x}$, we have that $\nabla g_I(\bar{x})z > 0$, and the weak Arrow-Hurwicz-Uzawa constraint qualification *3* is satisfied at $\bar{x}$.   ∎

We summarize the relations obtained above between the various constraint qualifications in Fig. *10.2.1*.

7   Kuhn-Tucker stationary-point necessary optimality theorem

Let X^0 be an open set in R^n, let θ and g be defined on X^0, let $\bar{x}$ solve LMP 7.1.2 or MP 7.1.1, let θ and g be differentiable at $\bar{x}$, and let g satisfy

(i)   *The Kuhn-Tucker constraint qualification 7.3.3 at $\bar{x}$, or*
(ii)   *the weak Arrow-Hurwicz-Uzawa constraint qualification 3 at $\bar{x}$, or*
(iii) *the weak reverse convex constraint qualification 4 at $\bar{x}$, or*
(iv)   *Slater's weak constraint qualification 5 at $\bar{x}$.*

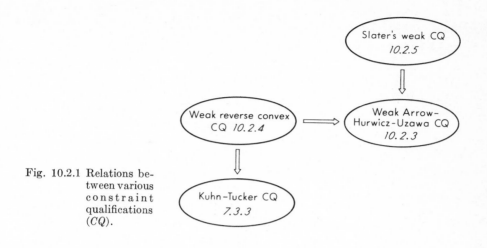

Fig. 10.2.1 Relations be-
        tween various
        constraint
        qualifications
        (CQ).

Then, there exists a $\bar{u} \in R^m$ such that $(\bar{x}, \bar{u})$ solves KTP 7.1.4.

PROOF   In view of Lemma 6 we need only establish the theorem under
assumptions (i) and (ii) above.

   (i)   This is the same as 7.3.7(i).

   (ii)   This proof is identical with the proof of Theorem 7.3.7(ii)
with the following modifications: V and W are replaced by P and Q, and
Theorem *2* is used in the proof instead of Theorem 7.3.2.   ∎

3.   Duality

In this section we will extend the Hanson-Huard strict converse duality
theorem *8.1.6* in two directions.   In the first direction we will show that
$(\hat{x}, \hat{u})$ need only be a local solution of the dual problem rather than a global
solution.   In the second direction we will relax the convexity of the objective function θ to pseudoconvexity, and the convexity of the constraint
function g to quasiconvexity.   We will also show that neither the weak
duality theorem *8.1.3* nor Wolfe's duality theorem *8.1.4* holds for a
pseudoconvex θ.

1   **Strict converse duality theorem [Mangasarian 65]**

         *Let X^0 be an open set in R^n, and let θ and g be differentiable on X^0.
Let $(\hat{x}, \hat{u})$ be a local solution of DP 8.1.2, that is,*

$$\psi(\hat{x}, \hat{u}) = \max_{(x,u) \in Y \cap B_\delta(\hat{x}, \hat{u})} \psi(x,u) \qquad (\hat{x}, \hat{u}) \in Y \cap B_\delta(\hat{x}, \hat{u})$$

where $B_\delta(\hat{x},\hat{u})$ is an open ball in $R^n \times R^m$ with radius $\delta > 0$ around $(\hat{x},\hat{u})$. Let θ be pseudoconvex at $\hat{x}$, and let g be quasiconvex at $\hat{x}$. If either

(i)   *$\psi(x,\hat{u})$ is twice continuously differentiable at $\hat{x}$, and the $n \times n$ Hessian matrix $\nabla_x^2\psi(\hat{x},\hat{u})$ is nonsingular, or*
(ii) *there exists an open set Λ in R^m containing $\hat{u}$ and an n-dimensional differentiable vector function e on Λ such that*

$$\hat{x} = e(\hat{u})$$

and

$$\left\langle \ e(u) \in X^0, \qquad \nabla_x\psi(x,u)\Big|_{x=e(u)} = 0 \ \right\rangle \qquad for \ u \in \Lambda$$

then $\hat{x}$ solves MP 8.1.1, and

$$\theta(\hat{x}) = \psi(\hat{x},\hat{u})$$

If $\bar{x}$ solves MP 8.1.1, it does not follow that $\bar{x}$ and some $\bar{u}$ solve DP 8.1.2 (even if g is linear), nor does it follow that $\theta(\bar{x}) \geq \psi(\bar{x},\bar{u})$ for any dual feasible point $(\bar{x},\bar{u})$, that is, $(\bar{x},\bar{u}) \in Y$.

PROOF   The proof of the first part of the theorem is essentially identical to the proof of Theorem *8.1.6*, except that we invoke the sufficient optimality theorem *10.1.2* here, whereas we invoked the sufficient optimality theorem *7.2.1* in the proof of *8.1.6*. (The assumption that $(\hat{x},\hat{u})$ need only be a local solution of DP *8.1.2* does not change the proof of *8.1.6* and in fact could have been made in Theorem *8.1.6* also. It would have added nothing to that theorem, however, since under the assumptions of Theorem *8.1.6* each local maximum is also a global maximum. This, however, is not the case under the present assumptions, as can be seen from the example following this proof, in which a local maximum to the dual problem is not a global maximum.)

We establish now the second part of the theorem by means of the following counterexample:

$$\min_{x \in X}\ [-e^{-(x)^2}] \qquad X = \{x \mid x \in R,\ -x + 1 \leq 0\} \tag{MP1}$$

$$\max_{(x,u) \in Y}\ (-e^{-(x)^2} - ux + u) \qquad Y = \left\{(x,u) \ \middle| \ \begin{array}{l} x \in R,\ u \in R \\[4pt] 2xe^{-(x)^2} - u = 0 \\[4pt] u \geq 0 \end{array} \right\} \tag{DP1}$$

The solution of MP1 is obviously $\bar{x} = 1$, whereas DP1 has no maximum but has a supremum, which is equal to zero. (To see this, rewrite the dual problem in the equivalent form

$$\max_{x \in R} \ \{-[1 - 2x + 2(x)^2]e^{-(x)^2} \mid x \geq 0\}$$

and note that the quadratic equation $1 - 2x + 2(x)^2 = 0$ has only complex roots.) We also have dual feasible points $(\bar{x}, \bar{u})$, such as $\bar{x} = 10$, $\bar{u} = 20e^{-100}$, such that $\theta(\bar{x}) < \psi(\bar{x}, \bar{u})$. ∎

We give now an example where the first part of the above theorem applies and where the dual problem has no global maximum but a local maximum which is also a solution of the primal problem. Consider the primal minimization problem

$$\min_{x \in X} x + (x)^3 \qquad X = \{x \mid x \in R, \ -(x)^3 - 1 \leq 0\} \qquad \text{(MP2)}$$

and its dual

$$\max_{(x,u) \in Y} x + (x)^3 - u(x)^3 - u \qquad Y = \left\{ (x,u) \ \middle| \ \begin{array}{l} x \in R, \ u \in R \\ 1 + 3(x)^2 - 3u(x)^2 = 0 \\ u \geq 0 \end{array} \right\}$$

$$\text{(DP2)}$$

Setting $v = 1 - u$ gives

$$\max_{(x,v)} \left\{ x + v(x)^3 + v - 1 \ \middle| \ \begin{array}{l} x \in R, \ v \in R \\ 1 + 3(x)^2 v = 0 \\ v \leq 1 \end{array} \right\}$$

and setting $v = -[1/3(x)^2]$, $x \neq 0$, gives

$$\max_{x \in R} \left\{ \frac{2x}{3} - \frac{1}{3(x)^2} - 1 \ \middle| \ x > 0 \text{ or } x < 0 \right\}$$

If we let $\phi(x) = (2/3)x - [1/3(x)^2] - 1$, we have that

$$\nabla\phi(x) = \frac{2}{3} + \frac{2}{3(x)^3} \qquad \text{and} \qquad \nabla^2\phi(x) = \frac{-2}{(x)^4} < 0$$

Hence $\nabla\phi(\hat{x}) = 0$ implies that $\hat{x} = -1$, and since $\nabla^2\phi(x) < 0$ for $x < 0$, ϕ is strictly concave on $\{x \mid x \in R, \ x < 0\}$, and $\hat{x} = -1$ is a local maximum of ϕ. However $\hat{x} = -1$ is not a global maximum of ϕ on $\{x \mid x \in R, x > 0 \text{ or } x < 0\}$, since φ approaches ∞ as x approaches ∞. However $(\hat{x}, \hat{u}) = (-1, 2/3)$ is a local solution of the dual problem and $\psi(\hat{x}, \hat{u}) = -2$.

It is obvious also that $\bar{x} = -1$ solves the primal problem and that $\theta(\bar{x}) = -2$. Hence $\bar{x} = \hat{x}$, and $\theta(\bar{x}) = \psi(\hat{x}, \hat{u})$.

We end this chapter with a corollary that follows directly from the proof of Theorem *8.1.6* or from the proof of Theorem *1* above.

2   ## Corollary

Let all the assumptions of Theorem 1 above hold, except that θ need not be pseudoconvex at x̂ nor g be quasiconvex at x̂. If (x̂,û) is a local solution of DP 8.1.2, then (x̂,û) is a Kuhn-Tucker point, that is, (x̂,û) solves KTP 7.1.4.

Chapter Eleven

Optimality and Duality in the Presence of Nonlinear Equality Constraints

Up until now we have dealt almost exclusively with minimization problems in which the feasible region X is determined by inequality constraints only, that is,

$$X = \{x \mid x \in X^0, g(x) \leq 0\}$$

where X^0 is a subset of R^n, and g is an m-dimensional vector function defined on X^0. (For exceptions see *5.3.2*, *5.4.8*, *7.2.2*, and *7.3.12*, where equality constraints are also considered.) In this chapter we shall examine in some detail minimization problems in which the feasible region X is determined by inequality and nonlinear equality constraints, that is,

$$X = \{x \mid x \in X^0, g(x) \leq 0,\\ h(x) = 0\}$$

where h is a k-dimensional vector function defined on X^0. In particular we shall obtain sufficient optimality criteria, necessary optimality criteria, and some duality results for the following minimization problem

$$\theta(\bar{x}) = \min_{x \in X} \theta(x)\\ \bar{x} \in X = \{x \mid x \in X^0,\\ g(x) \leq 0, h(x) = 0\}$$

We remark here that all the necessary optimality conditions of Chap. 7 that involved differentiable functions required that the set X^0 be open. We derive in this chapter an important necessary optimality condition of the maximum-principle type which does not require that X^0 be open.

1. Sufficient optimality criteria

The sufficient optimality criteria given here follow directly from Theorem 10.1.1 and Theorem 10.1.2 by observing that the equality constraint $h(x) = 0$ can be written as $h(x) \leq 0$ and $-h(x) \leq 0$, and that the negative of a quasiconcave function is quasiconvex.

1 Sufficient optimality theorem

Let X^0 be an open set in R^n, and let θ, g, and h be respectively a numerical function, an m-dimensional vector function, and a k-dimensional vector function, all defined on X^0. Let $\bar{x} \in X^0$, let

$$I = \{i \mid g_i(\bar{x}) = 0\}$$

let θ be pseudoconvex at $\bar{x}$, let g_I be differentiable and quasiconvex at $\bar{x}$, and let h be differentiable, quasiconvex, and quasiconcave at $\bar{x}$. If there exist $\bar{u} \in R^m$ and $\bar{v} \in R^k$ such that $(\bar{x}, \bar{u}, \bar{v})$ satisfies the following conditions

$$[\nabla\theta(\bar{x}) + \bar{u}\nabla g(\bar{x}) + \bar{v}\nabla h(\bar{x})](x - \bar{x}) \geq 0 \qquad \text{for all } x \in X$$

$$\bar{u}g(\bar{x}) = 0$$

$$g(\bar{x}) \leq 0$$

$$h(\bar{x}) = 0$$

$$\bar{u} \geq 0$$

then $\bar{x}$ is a solution of the following minimization problem

$$\theta(\bar{x}) = \min_{x \in X} \theta(x) \qquad \bar{x} \in X = \{x \mid x \in X^0, g(x) \leq 0, h(x) = 0\}$$

Kuhn-Tucker sufficient optimality theorem

Theorem 1 above holds with the condition

$$\nabla\theta(\bar{x}) + \bar{u}\nabla g(\bar{x}) + \bar{v}\nabla h(\bar{x}) = 0$$

replacing the condition

$$[\nabla\theta(\bar{x}) + \bar{u}\nabla g(\bar{x}) + \bar{v}\nabla h(\bar{x})](x - \bar{x}) \geq 0 \qquad \text{for all } x \in X$$

2. "Minimum-principle" necessary optimality criteria: X^0 not open

We consider in this section the minimization problem

$$\theta(\bar{x}) = \min_{x \in X} \theta(x) \qquad \bar{x} \in X = \{x \mid x \in X^0, g(x) \leq 0, h(x) = 0\}$$

We do not assume here that X^0 is an open subset of R^n, that g is convex on X^0, or that h is linear on R^n.[†] Without the assumption that X^0 is open, we cannot derive necessary optimality criteria of either the Fritz John type (7.3.2) or of the Kuhn-Tucker type (7.3.7). We can, however, derive necessary optimality criteria of the minimum-principle type, which, when X^0 is open, become necessary optimality criteria of the Fritz John or Kuhn-Tucker type. (The minimum-principle necessary optimality criterion we are about to derive has been referred to as a maximum principle in [Halkin 66, Halkin-Neustadt 66, Canon et al. 66, Mangasarian-Fromovitz 67] because of its close relation to Pontryagin's maximum principle [Pontryagin et al. 62], which is the fundamental optimality condition of optimal control. We have a minimum principle here instead of a maximum principle because the adjoint variables of Pontryagin's maximum principle, which play the role of Lagrange multipliers, are the negative of our Lagrange multipliers $\bar{r}_0$, $\bar{r}$, $\bar{s}$, to be introduced below. Halkin [Halkin 66] has a maximum principle because he has a maximization problem instead of a minimization problem.)

We point out now the main difference between a minimum-principle necessary optimality criterion and the main condition of the Fritz John necessary optimality criterion. The minimum principle requires that

$$[\bar{r}_0 \nabla \theta(\bar{x}) + \bar{r} \nabla g(\bar{x}) + \bar{s} \nabla h(\bar{x})](x - \bar{x}) \geqq 0 \qquad \text{for all } x \in \overline{X}^0$$

where $\bar{x}$ is a solution of the minimization problem, $\bar{r}_0$, $\bar{r}$, and $\bar{s}$ are Lagrange multipliers and $\overline{X}^0$ is the closure of X^0. The Fritz John criterion requires that

$$\bar{r}_0 \nabla \theta(\bar{x}) + \bar{r} \nabla g(\bar{x}) + \bar{s} \nabla h(\bar{x}) = 0$$

Obviously, the latter condition follows from the first one if X^0 is open. If X^0 is not open, the latter condition does not hold in general.

We begin by establishing a generalization of a linearization lemma of [Mangasarian-Fromovitz 67a].

1 Linearization lemma

Let X^0 be a convex set in R^n with a nonempty interior, int (X^0), and let Λ be an open set in R^n. Let f be an ℓ-dimensional vector function, and let h be a k-dimensional vector function, both defined on some open set containing X^0. Let

$$\bar{x} \in X^0 \cap \Lambda \qquad f(\bar{x}) = 0 \qquad h(\bar{x}) = 0$$

[†] For the case when X^0 is convex, g is convex on X^0, and h is linear on R^n, we have the necessary optimality conditions 5.4.8.

Let f be differentiable at $\bar{x}$, let h have continuous first partial derivatives at $\bar{x}$, and let $\nabla h_j(\bar{x})$, $j = 1, \ldots, k$, be linearly independent, that is,

$$\left\langle \begin{array}{l} y \in R^k \\ y\nabla h(\bar{x}) = 0 \end{array} \right\rangle \Rightarrow y = 0$$

Then

$$\left\langle \begin{array}{ll} f(x) < 0 & \text{has no solution} \\ h(x) = 0 & x \in X^0 \cap \Lambda \end{array} \right\rangle$$

$$\Rightarrow \left\langle \begin{array}{ll} \nabla f(\bar{x})(x - \bar{x}) < 0 & \text{has no solution} \\ \nabla h(\bar{x})(x - \bar{x}) = 0 & x \in \text{int } (X^0) \end{array} \right\rangle$$

PROOF $k = n$: The linear independence of $\nabla h_j(\bar{x})$, $j = 1, \ldots, k$, is equivalent to the nonsingularity of $\nabla h(\bar{x})$. Hence $\nabla f(\bar{x})(x - \bar{x}) < 0$ cannot hold, because $\nabla h(\bar{x})(x - \bar{x}) = 0$ implies that $x - \bar{x} = 0$.

$k > n$: This case is excluded because of the assumption that $\nabla h_j(\bar{x})$, $j = 1, \ldots, k$, are linearly independent, which implies that $k \leq n$.

$0 < k < n$: We shall establish the lemma by proving the following backward implication

$$\left\langle \begin{array}{ll} f(x) < 0 & \text{has a solution} \\ h(x) = 0 & \tilde{x} \in X^0 \cap \Lambda \end{array} \right\rangle$$

$$\Leftarrow \left\langle \begin{array}{ll} \nabla f(\bar{x})(x - \bar{x}) < 0 & \text{has a solution} \\ \nabla h(\bar{x})(x - \bar{x}) = 0 & \hat{x} \in \text{int } (X^0) \end{array} \right\rangle$$

which is the logical equivalent of the implication of the lemma.

Since $h(\bar{x}) = 0$, and since $\nabla h_j(\bar{x})$, $j = 1, \ldots, k$, are linearly independent, we have, by the implicit function theorem *D.3.1*, that there exist: (i) a partition (x_I, x_K) of x such that $x_I \in R^{n-k}$, $x_K \in R^k$, (ii) an open set Ω in R^{n-k} containing $\bar{x}_I$, and (iii) a k-dimensional differentiable vector function e on Ω such that

$$\bar{x}_K = e(\bar{x}_I)$$

$$h[x_I, e(x_I)] = 0 \qquad \text{for all } x_I \in \Omega$$

and

$\nabla_{x_K} h(\bar{x})$ is nonsingular

Since h and e are differentiable, and since $h[x_I, e(x_I)] = 0$ for all $x_I \in \Omega$, we have, by the chain rule $D.1.6$, that

$\nabla_{x_I} h(\bar{x}) + \nabla_{x_K} h(\bar{x}) \nabla e(\bar{x}_I) = 0$

Postmultiplication by $(\hat{x}_I - \bar{x}_I)$ gives

$\nabla_{x_I} h(\bar{x})(\hat{x}_I - \bar{x}_I) + \nabla_{x_K} h(\bar{x}) \nabla e(\bar{x}_I)(\hat{x}_I - \bar{x}_I) = 0$

But by assumption $\nabla h(\bar{x})(\hat{x} - \bar{x}) = 0$, or equivalently

$\nabla_{x_I} h(\bar{x})(\hat{x}_I - \bar{x}_I) + \nabla_{x_K} h(\bar{x})(\hat{x}_K - \bar{x}_K) = 0$

Hence the last two distinct equalities and the nonsingularity of $\nabla_{x_K} h(\bar{x})$ imply that

$\hat{x}_K - \bar{x}_K = \nabla e(\bar{x}_I)(\hat{x}_I - \bar{x}_I)$

Now, since Ω is open, since $\bar{x}_I \in \Omega$, and since e is differentiable at $\bar{x}$, there exists a $\delta_0 > 0$ such that for $\delta < \delta_0$

$\bar{x}_I + \delta(\hat{x}_I - \bar{x}_I) \in \Omega$

and

$e[\bar{x}_I + \delta(\hat{x}_I - \bar{x}_I)] = e(\bar{x}_I) + \delta \nabla e(\bar{x}_I)(\hat{x}_I - \bar{x}_I)$
$$+ \delta c[\bar{x}_I, \delta(\hat{x}_I - \bar{x}_I)] \|\hat{x}_I - \bar{x}_I\|$$

where c is a k-dimensional vector function that tends to zero as δ tends to zero. By combining the last two equalities and the fact that $\bar{x}_K = e(\bar{x}_I)$, we have that for $\delta < \delta_0$

$e[\bar{x}_I + \delta(\hat{x}_I - \bar{x}_I)] = \bar{x}_K + \delta(\hat{x}_K - \bar{x}_K) + \delta c[\bar{x}_I, \delta(\hat{x}_I - \bar{x}_I)] \|\hat{x}_I - \bar{x}_I\|$

But this relation and the fact that f is differentiable at $\bar{x}$ imply that there exists a δ_1 such that $0 < \delta_1 < \delta_0$ and such that for $\delta < \delta_1$

$f\{\bar{x}_I + \delta(\hat{x}_I - \bar{x}_I), e[\bar{x}_I + \delta(\hat{x}_I - \bar{x}_I)]\}$

$= f[\bar{x}_I + \delta(\hat{x}_I - \bar{x}_I), \bar{x}_K + \delta(\hat{x}_K - \bar{x}_K) + \delta c(\bar{x}_I, \delta(\hat{x}_I - \bar{x}_I)) \|\hat{x}_I - \bar{x}_I\|]$

$= f(\bar{x}) + \delta\{\nabla_{x_I} f(\bar{x})(\hat{x}_I - \bar{x}_I) + \nabla_{x_K} f(\bar{x})[(\hat{x}_K - \bar{x}_K)$

$+ c(\bar{x}_I, \delta(\hat{x}_I - \bar{x}_I)) \|\hat{x}_I - \bar{x}_I\|] + b[\bar{x}; \delta(\hat{x}_I - \bar{x}_I), \delta(\hat{x}_K - \bar{x}_K)$

$+ \delta c(\bar{x}_I, \delta(\hat{x}_I - \bar{x}_I)) \|\hat{x}_I - \bar{x}_I\|] \cdot \|[(\hat{x}_I - \bar{x}_I), (\hat{x}_K - \bar{x}_K)$

$+ c(\bar{x}_I, \delta(\hat{x}_I - \bar{x}_I)) \|\hat{x}_I - \bar{x}_I\|]\|\}$

where $\|x_I, x_K\|$ denotes $(x_I x_I + x_K x_K)^{1/2}$, and where b is an ℓ-dimensional vector function which tends to zero as δ tends to zero. Since c tends to zero as δ tends to zero, and since (by assumption) $\nabla f(\bar{x})(\hat{x} - \bar{x}) < 0$, that is,

$$\nabla_{x_I} f(\bar{x})(\hat{x}_I - \bar{x}_I) + \nabla_{x_K} f(\bar{x})(\hat{x}_K - \bar{x}_K) < 0$$

it follows that there exists a δ_2 such that $0 < \delta_2 < \delta_1$, and such that for $0 < \delta < \delta_2$, the quantity in the curly brackets { } in the expression next to the last is strictly negative. Hence (since $f(\bar{x}) = 0$) we have that

$$f\{\bar{x}_I + \delta(\hat{x}_I - \bar{x}_I), \, e[\bar{x}_I + \delta(\hat{x}_I - \bar{x}_I)]\} < 0 \qquad \text{for } 0 < \delta < \delta_2$$

and, since $h[x_I, e(x_I)] = 0$ for $x_I \in \Omega$, we also have that

$$h\{\bar{x}_I + \delta(\hat{x}_I - \bar{x}_I), \, e[\bar{x}_I + \delta(\hat{x}_I - \bar{x}_I)]\} = 0 \qquad \text{for } \delta < \delta_2$$

Now, we have already shown that for $\delta < \delta_0$

$$e[\bar{x}_I + \delta(\hat{x}_I - \bar{x}_I)] = (1 - \delta)\bar{x}_K + \delta\{\hat{x}_K + c[\bar{x}_I, \, \delta(\hat{x}_I - \bar{x}_I)]\|\hat{x}_I - \bar{x}_I\|\}$$

Because $\hat{x} \in \text{int}(X^0)$, and because c tends to zero as δ tends to zero, there exists a δ_3 such that $0 < \delta_3 < \delta_2$ and $\delta_3 < 1$, and such that

$$\{\hat{x}_I, \, \hat{x}_K + c[\bar{x}_I, \, \delta(\hat{x}_I - \bar{x}_I)]\|\hat{x}_I - \bar{x}_I\|\} \in X^0 \qquad \text{for } \delta < \delta_3$$

Since $\bar{x} \in X^0$, it follows from the last two relations and the convexity of X^0 that

$$\{\bar{x}_I + \delta(\hat{x}_I - \bar{x}_I), \, e[\bar{x}_I + \delta(\hat{x}_I - \bar{x}_I)]\} \in X^0 \qquad \text{for } \delta < \delta_3$$

Since $\bar{x} \in \Lambda$, and since Λ is open, it follows from the expression established above for $e[\bar{x}_I + \delta(\hat{x}_I - \bar{x}_I)]$ that there exists a δ_4 such that $0 < \delta_4 < \delta_3$ and such that

$$\{\bar{x}_I + \delta(\hat{x}_I - \bar{x}_I), \, e[\bar{x}_I + \delta(\hat{x}_I - \bar{x}_I)]\} \in \Lambda \qquad \text{for } \delta < \delta_4$$

Hence, combining the above results, we have that any $\tilde{x}$ given by

$$\tilde{x} = \{\bar{x}_I + \delta(\hat{x}_I - \bar{x}_I), \, e[\bar{x}_I + \delta(\hat{x}_I - \bar{x}_I)]\} \qquad 0 < \delta < \delta_4$$

satisfies $f(\tilde{x}) < 0$, $h(\tilde{x}) = 0$ and is in $X^0 \cap \Lambda$. We have thus established the lemma for the case when $n > k > 0$.

$k = 0$: This is the case where there is no $h(x) = 0$. We prove here that

$$f(\tilde{x}) < 0, \, \tilde{x} \in X^0 \cap \Lambda \Leftarrow \nabla f(\bar{x})(\hat{x} - \bar{x}) < 0, \, \hat{x} \in \text{int}(X^0)$$

We have now that for $0 < \delta < \delta_0$

$$f[\bar{x} + \delta(\hat{x} - \bar{x})] = f(\bar{x}) + \delta\{\nabla f(\bar{x})(\hat{x} - \bar{x}) + b[\bar{x}, \, \delta(\hat{x} - \bar{x})]\|\hat{x} - \bar{x}\|\}$$

Since b tends to zero as δ tends to zero, and since $\nabla f(\bar{x})(\hat{x} - \bar{x}) < 0$, there exists a $\hat{\delta} > 0$ such that

$$f[\bar{x} + \delta(\hat{x} - \bar{x})] < 0 \qquad \text{for } 0 < \delta < \hat{\delta}$$

Since $\bar{x} \in X^0 \cap \Lambda$, and since X^0 is convex and Λ is open, there exists a $\tilde{\delta} < 0$ such that $\tilde{\delta} < \hat{\delta}$ and $\tilde{\delta} < 1$, and such that

$$\bar{x} + \delta(\hat{x} - \bar{x}) \in X^0 \cap \Lambda \qquad \text{for } \delta < \tilde{\delta}$$

Hence any $\tilde{x} = \bar{x} + \delta(\hat{x} - \bar{x})$, $0 < \delta < \tilde{\delta}$, satisfies $f(\tilde{x}) < 0$ and is in $X^0 \cap \Lambda$. ∎

We derive now another lemma from the above one by using the fundamental theorem for convex functions, Theorem *4.2.1* (which is a consequence of the separation theorem for convex sets, Theorem *3.2.3*). This will be the key lemma in deriving the minimum-principle necessary optimality criterion.

2   **Lemma**

Let X^0 be a convex set in R^n with a nonempty interior, int (X^0), and let Λ be an open set in R^n. Let f be an ℓ-dimensional vector function, and let h be a k-dimensional vector function, both defined on some open set containing X^0. Let

$$\bar{x} \in X^0 \cap \Lambda \qquad f(\bar{x}) = 0 \qquad h(\bar{x}) = 0$$

Let f be differentiable at $\bar{x}$, and let h have continuous first partial derivatives at $\bar{x}$. Then

$$\left\langle \begin{array}{ll} f(x) < 0 & \text{has no} \\ & \text{solution} \\ h(x) = 0 & x \in X^0 \cap \Lambda \end{array} \right\rangle \Rightarrow \left\langle \begin{array}{l} \exists \bar{p} \in R^\ell,\ \bar{q} \in R^k \\ \bar{p} \geq 0,\ (\bar{p},\bar{q}) \neq 0: \\ [\bar{p}\nabla f(\bar{x}) + \bar{q}\nabla h(\bar{x})](x - \bar{x}) \geq 0 \\ \qquad\qquad \text{for all } x \in \bar{X}^0 \end{array} \right\rangle$$

PROOF If $\nabla h_j(\bar{x})$, $j = 1, \ldots, k$, are linearly dependent, then there exists a $\bar{q} \neq 0$ such that $\bar{q}\nabla h(\bar{x}) = 0$, and the lemma is trivially satisfied by $\bar{p} = 0$, $\bar{q} \neq 0$, and $\bar{q}\nabla h(\bar{x})(x - \bar{x}) = 0$ for all $x \in \bar{X}^0$.

If $\nabla h_j(\bar{x}), j = 1, \ldots, k$, are linearly independent, then by Lemma *1* above we have that

$$\left\langle \begin{array}{l} \nabla f(\bar{x})(x - \bar{x}) < 0 \\ \nabla h(\bar{x})(x - \bar{x}) = 0 \end{array} \right\rangle \text{ has no solution } x \in \text{int } (X^0)$$

Since the interior, int (X^0), of a convex set is convex (see *3.1.7*), it follows by Theorem *4.2.1* that there exist $\bar{p} \in R^l$, $\bar{q} \in R^k$, $\bar{p} \geq 0$, $(\bar{p},\bar{q}) \neq 0$ such that

$$[\bar{p}\nabla f(\bar{x}) + \bar{q}\nabla h(\bar{x})](x - \bar{x}) \geq 0 \qquad \text{for all } x \in X^0$$

and since the above expression is continuous in x, and in fact linear, the above inequality holds also on the closure $\bar{X}^0$ of X^0. ∎

We are now ready to derive the fundamental necessary optimality criterion of this chapter.

3   **Minimum-principle necessary optimality theorem**

Let X^0 be a convex set in R^n with a nonempty interior: int (X^0). *Let θ be a numerical function, let g be an m-dimensional vector function, and let h be a k-dimensional vector function, all defined on some open set containing X^0. Let $\bar{x}$ be a solution of*

$$\theta(\bar{x}) = \min_{x \in X} \theta(x) \qquad \bar{x} \in X = \{x \mid x \in X^0, g(x) \leq 0, h(x) = 0\}$$

Let θ and g be differentiable at $\bar{x}$, and let h have continuous first partial derivatives at $\bar{x}$. Then there exist $\bar{r}_0 \in R$, $\bar{r} \in R^m$, $\bar{s} \in R^k$ such that the following conditions are satisfied

$$[\bar{r}_0\nabla\theta(\bar{x}) + \bar{r}\nabla g(\bar{x}) + \bar{s}\nabla h(\bar{x})](x - \bar{x}) \geq 0 \qquad \text{for all } x \in \bar{X}^0$$

$$\bar{r}g(\bar{x}) = 0$$

$$(\bar{r}_0,\bar{r}) \geq 0$$

$$(\bar{r}_0,\bar{r},\bar{s}) \neq 0$$

REMARK The first of the above four relations is called a *minimum principle* because it is a necessary optimality condition for $\bar{r}_0\theta(x) + \bar{r}g(x) + \bar{s}h(x)$ to have a minimum at $\bar{x}$ (see Theorem *9.3.3*). If in addition we have that $\bar{r}_0\theta + \bar{r}g + \bar{s}h$ is pseudoconvex or convex at $\bar{x}$ (this, in general, is not the case under the assumptions of the above theorem), then the minimum-principle condition implies that

$$\bar{r}_0\theta(\bar{x}) + \bar{r}g(\bar{x}) + \bar{s}h(\bar{x}) = \min_{x \in X^0} \bar{r}_0\theta(x) + \bar{r}g(x) + \bar{s}h(x)$$

The above condition becomes then equivalent to the saddlepoint condition of Problem *5.4.2*.

PROOF Let

$$I = \{i \mid g_i(\bar{x}) = 0\} \qquad J = \{i \mid g_i(\bar{x}) < 0\} \qquad I \cup J = \{1,2, \ldots ,m\}$$

Let m_I and m_J denote the number of elements in the sets I and J respectively, so $m_I + m_J = m$.

Since g is defined on some open set containing X^0, and since g is differentiable at $\bar{x}$, there exists a $\bar{\delta} > 0$ such that for $i \in J$ and $\|x - \bar{x}\| < \bar{\delta}$

$$g_i(x) = g_i(\bar{x}) + \nabla g_i(\bar{x})(x - \bar{x}) + \alpha_i(\bar{x}, x - \bar{x})\|x - \bar{x}\|$$

$$\leqq g_i(\bar{x}) + [\|\nabla g_i(\bar{x})\| + \alpha_i(\bar{x}, x - \bar{x})]\|x - \bar{x}\| \qquad \text{(by 1.3.8)}$$

$$< 0 \qquad (\text{because } g_i(\bar{x}) < 0 \text{ and } \lim_{x - \bar{x} \to 0} \alpha_i(\bar{x}, x - \bar{x}) = 0)$$

Hence the set

$$\Lambda = \{x \mid g_J(x) < 0, \|x - \bar{x}\| < \bar{\delta}\}$$

is an open set in R^n.

Now, we have that the system

$$\left\langle \begin{array}{c} \theta(x) - \theta(\bar{x}) < 0 \\ g_I(x) < 0 \\ h(x) = 0 \end{array} \right\rangle \text{ has no solution } x \in X^0 \cap \Lambda$$

for if it did have a solution, then $\bar{x}$ would not be a solution of the minimization problem. But we also have that

$$\bar{x} \in X^0 \cap \Lambda \qquad \theta(\bar{x}) - \theta(\bar{x}) = 0 \qquad g_I(\bar{x}) = 0 \qquad h(\bar{x}) = 0$$

Hence by Lemma 2 above, there exist $\bar{r}_0 \in R$, $\bar{r}_I \in R^{m_I}$, $\bar{s} \in R^k$ such that

$$[\bar{r}_0 \nabla \theta(\bar{x}) + \bar{r}_I \nabla g_I(\bar{x}) + \bar{s} \nabla h(\bar{x})](x - \bar{x}) \geqq 0 \qquad \text{for all } x \in \bar{X}^0$$

$$(\bar{r}_0, \bar{r}_I) \geqq 0$$

$$(\bar{r}_0, \bar{r}_I, \bar{s}) \neq 0$$

By defining $\bar{r}_J = 0 \in E^{m_J}$ and observing that

$$\bar{r}g(\bar{x}) = \bar{r}_I g_I(\bar{x}) + \bar{r}_J g_J(\bar{x}) = 0$$

and

$$\bar{r}\nabla g(\bar{x}) = \bar{r}_I \nabla g_I(\bar{x}) + \bar{r}_J \nabla g_J(\bar{x}) = \bar{r}_I \nabla g_I(\bar{x})$$

the theorem is established. ∎

It should be remarked that the restriction that the convex set X^0 have a nonempty interior is a mild restriction, since in general a convex set without an interior is equivalent to the intersection of a convex set with a nonempty interior and a linear manifold $\{x \mid x \in R^n, h(x) = 0\}$ (h linear on R^n). Since we allow equalities $h(x) = 0$ in Theorem 3 above, convex sets with empty interiors can, in effect, be handled by the above results. The convexity requirement on X^0 is of course a restriction which cannot be dispensed with easily. (See however [Halkin 66, Canon et al. 66].) If we replace the convexity requirement on X^0 by the requirement that X^0 be open, a stronger necessary optimality condition than the above one can be obtained. In effect this will be an extension of the Fritz John stationary-point necessary optimality theorem 7.3.2 to the case of non-linear equalities. We shall give this result in the next section of this chapter.

3. Fritz John and Kuhn-Tucker stationary-point necessary optimality criteria: X^0 open

We derive in this section necessary optimality criteria of the Fritz John and Kuhn-Tucker types from the minimum principle of the previous section.

1   **Fritz John stationary-point necessary optimality theorem [Mangasarian-Fromovitz 67a]**

Let X^0 be an open set in R^n. Let θ be a numerical function on X^0, let g be an m-dimensional vector function on X^0, and let h be a k-dimensional vector function on X^0. Let $\bar{x}$ be a (global) solution of the minimization problem

$$\theta(\bar{x}) = \min_{x \in X} \theta(x), \quad \bar{x} \in X = \{x \mid x \in X^0, g(x) \leq 0, h(x) = 0\}$$

or a local solution thereof, that is,

$$\theta(\bar{x}) = \min_{x \in X \cap B_\delta(\bar{x})} \theta(x) \qquad \bar{x} \in X \cap B_\delta(\bar{x})$$

where $B_\delta(\bar{x})$ is an open ball around $\bar{x}$ with radius δ. Let θ and g be differentiable at $\bar{x}$, and let h have continuous first partial derivatives at $\bar{x}$. Then there exist $\bar{r}_0 \in R$, $\bar{r} \in R^m$, $\bar{s} \in R^k$ such that the following conditions are satisfied

$$\bar{r}_0 \nabla \theta(\bar{x}) + \bar{r} \nabla g(\bar{x}) + \bar{s} \nabla h(\bar{x}) = 0$$

$$\bar{r} g(\bar{x}) = 0$$

$$(\bar{r}_0, \bar{r}) \geq 0$$

$$(\bar{r}_0, \bar{r}, \bar{s}) \neq 0$$

PROOF Let $\bar{x}$ be a global or local solution of the minimization problem.

In either case (since X^0 is open) there exists an open ball $B_\rho(\bar{x})$ around $\bar{x}$ with radius ρ such that $B_\rho(\bar{x}) \subset B_\delta(\bar{x}) \subset X^0$, and

$$\theta(\bar{x}) = \min_{x \in X^*} \theta(x) \qquad \bar{x} \in X^* = \{x \mid x \in B_\rho(\bar{x}),\, g(x) \leqq 0,\, h(x) = 0\}$$

Since $B_\rho(\bar{x})$ is a convex set with a nonempty interior, we have by the minimum principle *11.2.3* that there exist $\bar{r}_0 \in R$, $\bar{r} \in R^m$, $\bar{s} \in R^k$ such that

$$[\bar{r}_0 \nabla \theta(\bar{x}) + \bar{r} \nabla g(\bar{x}) + \bar{s} \nabla h(\bar{x})](x - \bar{x}) \geqq 0 \qquad \text{for all } x \in B_\rho(\bar{x})$$

$$\bar{r}g(\bar{x}) = 0$$

$$(\bar{r}_0, \bar{r}) \geqq 0$$

$$(\bar{r}_0, \bar{r}, \bar{s}) \neq 0$$

Since for some small positive ζ

$$\bar{x} - \zeta[\bar{r}_0 \nabla \theta(\bar{x}) + \bar{r} \nabla g(\bar{x}) + \bar{s} \nabla h(\bar{x})] \in B_\rho(\bar{x})$$

we have from the first inequality above that

$$\bar{r}_0 \nabla \theta(\bar{x}) + \bar{r} \nabla g(\bar{x}) + \bar{s} \nabla h(\bar{x}) = 0 \quad \blacksquare$$

To derive Kuhn-Tucker conditions from the above, we need to impose constraint qualifications on the problem.

2   The Kuhn-Tucker constraint qualification (see *7.3.3*)

Let X^0 be an open set in R^n, let g and h be m-dimensional and k-dimensional vector functions on X^0, and let

$$X = \{x \mid x \in X^0,\, g(x) \leqq 0,\, h(x) = 0\}$$

g and h are said to satisfy the *Kuhn-Tucker constraint qualification at* $\bar{x} \in X$ if g and h are differentiable at $\bar{x}$ and

$$\left.\begin{array}{c} y \in R^n \\[4pt] \nabla g_I(\bar{x})y \leqq 0 \\[4pt] \nabla h(\bar{x})y = 0 \end{array}\right\} \Rightarrow \left\{\begin{array}{l} \text{There exists an } n\text{-dimensional vector function} \\ e \text{ on the interval } [0,1] \text{ such that} \\[4pt] a.\ e(0) = \bar{x} \\[4pt] b.\ e(\tau) \in X \text{ for } 0 \leqq \tau \leqq 1 \\[4pt] c.\ e \text{ is differentiable at } \tau = 0 \text{ and } \dfrac{de(0)}{d\tau} = \lambda y \\[4pt] \qquad\qquad\qquad \text{for some } \lambda > 0 \end{array}\right.$$

where

$$I = \{i \mid g_i(\bar{x}) = 0\}$$

3   **The weak Arrow-Hurwicz-Uzawa constraint qualification**
          (see *10.2.3*)

Let X^0 be an open set in R^n, let g and h be m-dimensional and k-dimensional vector functions on X^0, and let

$$X = \{x \mid x \in X^0, g(x) \leq 0, h(x) = 0\}$$

g and h are said to satisfy the *weak Arrow-Hurwicz-Uzawa constraint qualification at* $\bar{x} \in X$ if g and h are differentiable at $\bar{x}$, h is both pseudo-convex and pseudoconcave at $\bar{x}$, and

$$\left.\begin{array}{l} \nabla g_Q(\bar{x})z > 0 \\[4pt] \nabla g_P(\bar{x})z \geq 0 \\[4pt] \nabla h(\bar{x})z = 0 \end{array}\right\} \text{ has a solution } z \in R^n$$

where

$$P = \{i \mid g_i(\bar{x}) = 0, \text{ and } g_i \text{ is pseudoconcave at } \bar{x}\}$$

$$Q = \{i \mid g_i(\bar{x}) = 0, \text{ and } g_i \text{ is not pseudoconcave at } \bar{x}\}$$

4   **The weak reverse convex constraint qualification** (see *10.2.4*)

Let X^0 be an open set in R^n, let g and h be m-dimensional and k-dimensional vector functions on X^0, and let

$$X = \{x \mid x \in X^0, g(x) \leq 0, h(x) = 0\}$$

g and h are said to satisfy the *weak reverse constraint qualification at* $\bar{x} \in X$ if g and h are differentiable at $\bar{x}$, and if for each

$$i \in I = \{i \mid g_i(\bar{x}) = 0\}$$

either g_i is pseudoconcave at $\bar{x}$ or linear on R^n, and h is both pseudo-convex and pseudoconcave at $\bar{x}$.

5   **The modified Arrow-Hurwicz-Uzawa constraint qualification**
          [Mangasarian-Fromovitz 67a]

Let X^0 be an open set in R^n, let g and h be m-dimensional and k-dimensional vector functions on X^0, and let

$$X = \{x \mid x \in X^0, g(x) \leq 0, h(x) = 0\}$$

g and h are said to satisfy the *modified Arrow-Hurwicz-Uzawa constraint qualification at* $\bar{x} \in X$ if g is differentiable at $\bar{x}$, h is continuously differentiable at $\bar{x}$, $\nabla h_i(\bar{x})$, $i = 1, \ldots, k$, are linearly independent, and

$$\left\langle \begin{array}{c} \nabla g_I(\bar{x})z > 0 \\ \nabla h(\bar{x})z = 0 \end{array} \right\rangle \text{ has a solution } z \in R^n$$

where

$$I = \{i \mid g_i(\bar{x}) = 0\}$$

6   **Kuhn-Tucker stationary-point necessary optimality theorem**

Let X^0 be an open set in R^n, and let θ, g, and h be respectively a numerical function, an m-dimensional vector function, and a k-dimensional vector function, all defined on X^0. Let $\bar{x}$ be a (global) solution of the minimization problem

$$\theta(\bar{x}) = \min_{x \in X} \theta(x) \qquad \bar{x} \in X = \{x \mid x \in X^0, g(x) \leq 0, h(x) = 0\}$$

or a local solution thereof, that is,

$$\theta(\bar{x}) = \min_{x \in X \cap B_\delta(\bar{x})} \theta(x) \qquad \bar{x} \in X \cap B_\delta(\bar{x})$$

where $B_\delta(\bar{x})$ is some open ball around $\bar{x}$ with radius δ. Let θ, g, and h be differentiable at $\bar{x}$, and let g and h satisfy

(i)   *the Kuhn-Tucker constraint qualification 2 at $\bar{x}$, or*
(ii) *the weak Arrow-Hurwicz-Uzawa constraint qualification 3 at $\bar{x}$, or*
(iii) *the weak reverse convex constraint qualification 4 at $\bar{x}$, or*
(iv) *the modified Arrow-Hurwicz-Uzawa constraint qualification 5 at $\bar{x}$.*

Then there exist $\bar{u} \in R^m$ and a $\bar{v} \in R^k$ such that

$$\nabla \theta(\bar{x}) + \bar{u}\nabla g(\bar{x}) + \bar{v}\nabla h(\bar{x}) = 0$$

$$g(\bar{x}) \leq 0$$

$$h(\bar{x}) = 0$$

$$\bar{u}g(\bar{x}) = 0$$

$$\bar{u} \geq 0$$

PROOF (i)–(ii)–(iii)   These parts of the theorem follow from Theorem 10.2.7, parts (i), (ii), and (iii), by replacing $h(x) = 0$ in the above theorem by $h(x) \leq 0$ and $-h(x) \leq 0$.

(iv)   All we have to show here is that $\bar{r}_0 > 0$ in the Fritz John theorem 1 above, for then we have that $\bar{x},\bar{u} = \bar{r}/\bar{r}_0$ and $\bar{v} = \bar{s}/\bar{r}_0$ satisfy the Kuhn-Tucker conditions above.   We assume now that $\bar{r}_0 = 0$ and exhibit a contradiction.

If $I = \{i \mid g_i(\bar{x}) = 0\}$ is empty, that is, there are no active constraints, then $\bar{r} = 0$ (since $\bar{r}g(\bar{x}) = 0$, $\bar{r} \geq 0$ and $g(\bar{x}) \leq 0$).   Hence by Theorem 1

$$\bar{s}\nabla h(\bar{x}) = 0 \qquad \bar{s} \neq 0$$

which contradicts the assumption of 5 that $\nabla h_i(\bar{x})$, $i = 1, \ldots, k$, are linearly independent.   (If there are no equality constraints $h(x) = 0$, then $(\bar{r}_0,\bar{r}) = 0$ contradicts the condition $(\bar{r}_0,\bar{r}) \neq 0$ of Theorem 1, since there is no $\bar{s}$.)

If I is not empty, then by Theorem 1 we have that

$$\bar{r}_I\nabla g_I(\bar{x}) + \bar{s}\nabla h(\bar{x}) = 0$$

$$\bar{r}_I \geq 0$$

$$(\bar{r}_I,\bar{s}) \neq 0$$

If $\bar{r}_I = 0$, then $\bar{s} \neq 0$, and we have a contradiction to the assumption of 5 that $\nabla h_i(\bar{x})$, $i = 1, \ldots, k$, are linearly independent (if there is no $\bar{s}$, then $\bar{r}_I = 0$ implies $(\bar{r}_0,\bar{r}) = 0$, which contradicts the condition $(\bar{r}_0,\bar{r}) \neq 0$ of Theorem 1).   If $r_I \neq 0$, then $\bar{r}_I \geq 0$.   But by 5 there exists a z such that

$$\nabla g_I(\bar{x})z > 0 \qquad \text{and} \qquad \nabla h(\bar{x})z = 0$$

Hence

$$\bar{r}_I\nabla g_I(\bar{x})z + \bar{s}\nabla h(\bar{x})z > 0$$

which contradicts the equality above that

$$\bar{r}_I\nabla g_I(\bar{x}) + \bar{s}\nabla h(\bar{x}) = 0 \quad \blacksquare$$

4.   Duality with nonlinear equality constraints

The following strict converse duality results follow respectively from Theorem $10.3.1$ and Corollary $10.3.2$ by replacing the equality constraints $h(x) = 0$ by $h(x) \leq 0$ and $-h(x) \leq 0$.

1   **Strict converse duality theorem**

Let X^0 be an open set in R^n, and let θ, g, and h be respectively a numerical function, an m-dimensional vector function, and a k-dimensional vector function, all defined and differentiable on X^0.   Let $(\hat{x},\hat{u},\hat{v})$ be a (global) *solution of the dual problem*

$$
2 \quad \begin{cases}
\psi(\hat{x},\hat{u},\hat{v}) = \max_{(x,u,v)\in Y} \psi(x,u,v) \\[2mm]
\psi(x,u,v) = \theta(x) + ug(x) + vh(x) \\[2mm]
(\hat{x},\hat{u},\hat{v}) \in Y = \left\{ (x,u,v) \middle| \begin{array}{l} x \in X^0,\ u \in R^m,\ v \in R^k \\ \nabla_x\psi(x,u,v) = 0 \\ u \geqq 0 \end{array} \right\}
\end{cases}
$$

or a local solution of the dual problem, that is,

$$
\psi(\hat{x},\hat{u},\hat{v}) = \max_{(x,u,v)\in Y\cap B_\delta(\hat{x},\hat{u},\hat{v})} \psi(x,u,v)
$$

$$
(\hat{x},\hat{u},\hat{v}) \in Y \cap B_\delta(\hat{x},\hat{u},\hat{v})
$$

where $B_\delta(\hat{x},\hat{u},\hat{v})$ is an open ball in $R^n \times R^m \times R^k$ with radius $\delta > 0$ around $(\hat{x},\hat{u},\hat{v})$. Let θ be pseudoconvex at $\hat{x}$, let g be quasiconvex at $\hat{x}$, and let h be both quasiconvex and quasiconcave at $\hat{x}$. If either

(i) *$\psi(x,\hat{u},\hat{v})$ is twice continuously differentiable at $\hat{x}$ and the $n \times n$ Hessian matrix $\nabla_x^2\psi(\hat{x},\hat{u},\hat{v})$ is nonsingular, or*

(ii) *there exists an open set Λ in $R^m \times R^k$ containing $(\hat{u},\hat{v})$ and an n-dimensional differentiable vector function e on Λ such that*

$$
\hat{x} = e(\hat{u},\hat{v})
$$

and

$$
\left\langle e(u,v) \in X^0,\ \nabla_x\psi(x,u,v) \Big|_{x\,=\,e(u,v)} = 0 \right\rangle \qquad \textit{for } (u,v) \in \Lambda
$$

then $\hat{x}$ solves the minimization problem

$$
3 \quad \min_{x\in X} \theta(x) \qquad X = \{x \mid x \in X^0,\ g(x) \leqq 0,\ h(x) = 0\}
$$

and

$$
\theta(\hat{x}) = \psi(\hat{x},\hat{u},\hat{v})
$$

If $\bar{x}$ solves the minimization problem 3, it does not follow that $\bar{x}$ and some $\bar{u}$ and $\bar{v}$ solve the dual problem 2 (unless X^0 is convex, θ and g are convex on X^0, and h is linear on R^n), nor does it follow that $\theta(\bar{x}) \geqq \psi(\bar{x},\bar{u},\bar{v})$ for any dual feasible point $(\bar{x},\bar{u},\bar{v})$, that is, $(\bar{x},\bar{u},\bar{v}) \in Y$.

4   Corollary

     Let all the assumptions of Theorem 1 above hold, except that θ need not be pseudoconvex at $\hat{x}$, nor g quasiconvex at $\hat{x}$, nor h both quasiconvex and quasiconcave at $\hat{x}$. If $(\hat{x},\hat{u},\hat{v})$ is a local solution of the dual problem 2, then $(\hat{x},\hat{u},\hat{v})$ is a Kuhn-Tucker point, that is,

$$\nabla\theta(\hat{x}) + \hat{u}\nabla g(\hat{x}) + \hat{v}\nabla h(\hat{x}) = 0$$

$$g(\hat{x}) \leqq 0$$

$$h(\hat{x}) = 0$$

$$\hat{u}g(\hat{x}) = 0$$

$$\hat{u} \geqq 0$$

Appendix A

Vectors and Matrices

We collect in this appendix some of the bare essentials of real linear algebra which are needed in this book. For more detail the reader is referred to [Gale 60, Halmos 58]. We assume here that the reader is familiar with the notation and definitions of Chap. 1.

1. Vectors

Fundamental theorem of vector spaces

Let each of the vectors y^0, y^1, . . . , y^m in R^n be a linear combination of the vectors x^1, x^2, . . . , x^m in R^n; then y^0, y^1, . . . , y^m are linearly dependent.

PROOF The proof will be by induction on m. If $m = 1$, then $y^0 = p_0 x^1$, $y^1 = p_1 x^1$. If

$$p_0 = p_1 = 0$$

then $y^0 = y^1 = 0$, and y^0 and y^1 are linearly dependent. If not, then $p_0 \neq 0$, say. Then

$$p_1 y^0 - p_0 y^1 = p_1 p_0 x^1 - p_0 p_1 x^1 = 0$$

and y^0 and y^1 are linearly dependent because $p_0 \neq 0$.

Now assume that the theorem holds for $m = k - 1$, and we will show that it also holds for $m = k$. By hypothesis we have that

$$y^j = \sum_{i=1}^{k} p_i{}^j x^i \qquad j = 0, 1, \ldots, k$$

If all $p_i{}^j$ are zero then all y^j are zero and hence linearly dependent. Assume now that at least one $p_i{}^j$ is not zero, say $p_1{}^0$. Define

$$z^j = y^j - \frac{p_1{}^j}{p_1{}^0} y^0 = \sum_{i=2}^{k} \left(p_i{}^j - \frac{p_1{}^j}{p_1{}^0} p_i{}^0 \right) x^i \qquad j = 1, \ldots, k$$

Then each of the k vectors z^j is a linear combination of the $k - 1$ vectors $x^2, \ldots, x^k$, and hence by the induction hypothesis, the z^j are linearly dependent, that is, there exist numbers $q_1, \ldots, q_k$, not all zero, such that

$$0 = \sum_{j=1}^{k} q_j z^j = \sum_{j=1}^{k} q_j y^j - \frac{1}{p_1{}^0} \left(\sum_{j=1}^{k} q_j p_1{}^j \right) y^0$$

which shows that the y^j are linearly dependent.   ∎

2   **Corollary**

Any $n + 1$ vectors in R^n are linearly dependent.

PROOF   Let e^i be the vector in R^n with zero for each element except the ith, which is 1.   Then any vector x in R^n is a linear combination of the e^i for $x = \sum_{i=1}^{n} x_i e^i$.   The corollary then follows from Theorem *1* above.   ∎

3   **Corollary**

Any m vectors in R^n are linearly dependent if $m > n$.

PROOF   By Corollary *2* any $n + 1$ vectors of the m vectors are linearly dependent, hence the m vectors are linearly dependent.   ∎

4   **Corollary**

Each system of n homogeneous linear equations in m unknowns, $m > n$, has a nonzero solution.

PROOF   Consider the matrix equation $Ax = 0$ where A is any $n \times m$ matrix.   The m columns $A_{.j}$ of A are vectors in R^n and hence by Corollary *3* above are linearly dependent.   Hence there exists an $x \neq 0$ such that $Ax = 0$.   ∎

5   **Basis of a subset of R^n**

Let S be a subset of R^n, and let r be the maximum number of linearly independent vectors which can be chosen from S.   Any set of r linearly independent vectors of S is called a *basis for S*.

6   **Basis theorem**

The linearly independent vectors $x^1, \ldots, x^r$ are a basis for a subset S of R^n if and only if every vector y in S is a linear combination of the x^i.

PROOF [Gale 60] Suppose every y in S is a linear combination of the x^i. Then S contains no larger set of linearly independent vectors, for any set of more than r vectors must be dependent since they are combinations of the x^i. Therefore r is the maximum number of linearly independent vectors which can be chosen from S, and hence the x^i are a basis for S.

Conversely, suppose that the x^i are a basis for S. Then by definition, r is the number of vectors in the largest set of linearly independent vectors that can be found in S. So if y is in S, then $x^1, \ldots, x^r, y$ are linearly dependent, that is,

$$\sum_{i=1}^{r} p_i x^i + p_0 y = 0$$

and $p_0 \neq 0$, for otherwise the x^i would be linearly dependent. Hence

$$y = -\frac{1}{p_0} \sum_{i=1}^{r} p_i x^i$$

which is the desired linear combination. ∎

2.   Matrices

1   **Row and column rank of matrix**

The maximum number of linearly independent rows of a matrix is called the *row rank of the matrix,* and the maximum number of linearly independent columns of a matrix is called the *column rank of the matrix.*

2   **Rank theorem**

For any matrix A, the row rank and column rank are equal.

PROOF [Gale 60] Let r be the row rank of A, s its column rank, and suppose that $r < s$. Choose a row basis for A, which we may assume consists of the rows $A_1, \ldots, A_r$ (renumbering rows and columns of A clearly does not affect its row or column rank), and a column basis, which we assume consists of the columns $A_{.1}, \ldots, A_{.s}$. Let

$$\bar{A}_i = (A_{i1}, \ldots, A_{is}),$$

and note that the equations

$$\bar{A}_i y = 0 \qquad i = 1, \ldots, r$$

are r equations in s ($> r$) unknowns and hence have a nonzero solution $\bar{y}$ by Corollary $A.1.4$. Also since $A_1, \ldots, A_r$ are a row basis, it follows from the basis theorem $A.1.6$ that for all k, $A_k = \sum_{i=1}^{r} p_i{}^k A_i$ for some numbers $p_i{}^k$. Hence

$$\bar{A}_k = \sum_{i=1}^{r} p_i{}^k \bar{A}_i$$

and

$$\bar{A}_k \bar{y} = \sum_{i=1}^{r} p_i{}^k (\bar{A}_i \bar{y}) = 0 \qquad \text{for all } k$$

which is equivalent to

$$\sum_{j=1}^{s} \bar{A}_{.j} \bar{y}_j = 0$$

This shows that the columns $A_{.1}, \ldots, A_{.s}$ of A are linearly dependent, which contradicts the assumption that they were a basis. This contradiction shows that $r \geq s$. The same argument applied to the transpose of A shows that $s \geq r$. Hence $r = s$. ∎

3   **Rank of a matrix**

The *rank of a matrix* is the column or row rank of the matrix (which are equal by the above theorem).

4   **Corollary**

Let A be an $r \times n$ matrix with rank r ($r \leq n$). Then for any b in R^r, the system $Ax = b$ has a solution x in R^n.

PROOF   By Theorem 2 above, the column rank of A is r, and thus if $A_{.1}, \ldots, A_{.r}$, say, are a column basis, we have r linearly independent vectors in R^r, and the vector $b \in R^r$ can be expressed as a linear combination of them, that is,

$$b = \sum_{j=1}^{r} A_{.j} x_j$$

By defining $x_j = 0$ for $j = r + 1, \ldots, n$, we have that $b = Ax$. ∎

5   ### Nonsingular matrix

An $n \times n$ matrix is said to be *nonsingular* if it has rank n.

6   ### Semidefinite matrix

An $n \times n$ matrix A is said to be *positive semidefinite* if $xAx \geq 0$ for all x in R^n and *negative semidefinite* if $xAx \leq 0$ for all x in R^n.

7   ### Definite matrix

An $n \times n$ matrix A is said to be *positive definite* if

$$x \neq 0 \Rightarrow xAx > 0$$

and *negative definite* if

$$x \neq 0 \Rightarrow xAx < 0$$

Obviously the negative of a positive semidefinite (definite) matrix is a negative semidefinite (definite) matrix and conversely. Also, each positive (negative) definite matrix is also positive (negative) semidefinite.

8   ### Proposition

Each positive or negative definite matrix is nonsingular.

PROOF   Let A be an $n \times n$ positive or negative definite matrix. If A is singular, then its rank is less than n, and there exists an $x \in R^n$, $x \neq 0$, such that $Ax = 0$. Hence $xAx = 0$ for some $x \neq 0$, which contradicts the assumption that A is positive or negative definite. Hence A is nonsingular. ■

Appendix B

Résumé of Some Topological Properties of R^n

We collect here some well-known topological properties of R^n which can be found in greater detail in any modern book on real analysis or general topology [Buck 65, Fleming 65, Rudin 64, Berge 63, Simmons 63].

1. Open and closed sets

1 Open ball

Given a point $x^0 \in R^n$ and a real number $\lambda > 0$, the set

$$B_\lambda(x^0) = \{x \mid x \in R^n, \|x - x^0\| < \lambda\}$$

is called an *open ball* around x^0 with radius λ. (Sometimes the subscript λ from $B_\lambda(x^0)$ will be dropped for convenience.)

2 Interior point

A point x is said to be an *interior point* of the set $\Gamma \subset R^n$ if there exists a number $\epsilon > 0$ such that $B_\epsilon(x) \subset \Gamma$.

3 Point of closure

A point x is said to be a *point of closure* of the set $\Gamma \subset R^n$ if for each number $\epsilon > 0$,

$$B_\epsilon(x) \cap \Gamma \neq \emptyset$$

Note that a point of closure of Γ need not be in Γ. For example $0 \in R$ is a point of closure of the infinite set $\Gamma = \{1, \frac{1}{2}, \frac{1}{4}, \ldots\}$, but $0 \notin \Gamma$. On the other hand, every point in the set Γ is also a point of closure of Γ. In other words, a point of closure x^0 of a set Γ is a point such that there exist points in Γ that are arbitrarily close to it (closeness being measured by the distance $\delta(x, x^0) = \|x - x^0\|$, see *1.3.9*).

4   **Open set**

A set $\Gamma \subset R^n$ such that every point of Γ is an interior point is said to be *open*.

5   **Closed set**

A set $\Gamma \subset R^n$ such that every point of closure of Γ is in Γ is said to be *closed*.

6   **Closure of a set**

The *closure* $\bar{\Gamma}$ of a set $\Gamma \subset R^n$ is the set of points of closure of Γ. Obviously $\Gamma \subset \bar{\Gamma}$, and for a closed set $\Gamma = \bar{\Gamma}$.

7   **Interior of a set**

The *interior* int (Γ) of a set $\Gamma \subset R^n$ is the set of interior points of Γ. Obviously int $(\Gamma) \subset \Gamma$, and for an open set $\Gamma = $ int (Γ).

8   **Relatively open (closed) sets**

Let Γ and Λ be two sets such that $\Gamma \subset \Lambda \subset R^n$. Γ is said to be *open (closed) relative to* Λ if $\Gamma = \Lambda \cap \Omega$, where Ω is some open (closed) set in R^n.

Obviously an open (closed) set Γ in R^n is open (closed) relative to R^n. If $\Gamma \subset \Lambda$, and if Γ is open (closed), then Γ is open (closed) relative to Λ for $\Gamma = \Lambda \cap \Gamma$.

9   **Problem**

Show that:

(i)   Every open ball $B_\lambda(a)$ in R^n is an open set.
(ii) The closure $\overline{B_\lambda(a)}$ of an open ball $B_\lambda(a)$ in R^n is

$$\overline{B_\lambda(a)} = \{x \mid \|x - a\| \leq \lambda\}$$

and is closed. (The closure of an open ball is called a *closed ball* and is denoted by $\bar{B}_\lambda(a)$.)
(iii) The interior of a closed ball $\bar{B}_\lambda(a)$ is the open ball $B_\lambda(a)$.

10   **Theorem**

The family of open sets in R^n has the following properties:

(i)   *Every union of open sets is open.*
(ii) *Every finite intersection of open sets is open.*
(iii) *The empty set $\emptyset$ and R^n are open.*

PROOF (i) Let $(\Gamma_i)_{i \in I}$ be a family, finite or infinite, of open sets in R^n. If $x \in \Gamma = \bigcup_{i \in I} \Gamma_i$, then $x \in \Gamma_i$ for some $i \in I$, and there exists an $\epsilon > 0$ such that

$$B_\epsilon(x) \subset \Gamma_i \subset \Gamma$$

Hence x is an interior point of Γ, and Γ is open.

(ii) Let $\Gamma_{i \in I}$ be a finite family of open sets in R^n. If $x \in \Gamma = \bigcap_{i \in I} \Gamma_i$, then $x \in \Gamma_i$ for each $i \in I$. Because each Γ_i is open, there exist $\epsilon^i > 0$ such that

$$B_{\epsilon^i}(x) \subset \Gamma_i \qquad \text{for } \forall i \in I$$

Take $\epsilon = \underset{i}{\text{minimum }} \epsilon^i > 0$ (this is where the finiteness of I is used), then

$$B_\epsilon(x) \subset \Gamma$$

and Γ is open.

(iii) Since the empty set $\emptyset$ contains no points, we need not find an open ball surrounding any point, and hence $\emptyset$ is open. The set R^n is open because for each $x \in R^n$, $B_\epsilon(x) \subset R^n$ for all $\epsilon > 0$. ∎

The family of open sets in R^n, as defined in *4*, is called a *topology* in R^n. (In fact any family of sets—called open—which has properties (i), (ii), (iii) above is also called a topology in R^n. For example the sets $\emptyset$, R^n also form a topology in R^n. We shall, however, be concerned here only with open sets as defined by *4*.)

11   Theorem

Let $\Gamma \subset R^n$. Then

$$R^n \sim \bar{\Gamma} = \text{int } (R^n \sim \Gamma)$$

PROOF Let $x \in R^n$. Then

$$x \in \bar{\Gamma} \Leftrightarrow x \notin \text{int } (R^n \sim \Gamma)$$

Hence

$$x \notin \bar{\Gamma} \Leftrightarrow x \in \text{int } (R^n \sim \Gamma) \quad ∎$$

12   Corollary

The complement (relative to R^n) of an open set in R^n is closed, and vice versa.

PROOF Γ is closed $\Leftrightarrow \Gamma = \bar{\Gamma}$

$$\Leftrightarrow R^n \sim \Gamma = R^n \sim \bar{\Gamma} = \text{int} (R^n \sim \Gamma) \qquad \text{(by 11)}$$

$$\Leftrightarrow R^n \sim \Gamma \text{ is open} \quad \blacksquare$$

By using Corollary *12*, the following theorem is a direct consequence of Theorem *10*.

13 **Theorem**

The family of closed sets in R^n has the following properties:

(i) *Every intersection of closed sets is closed.*
(ii) *Every finite union of closed sets is closed.*
(iii) *The empty set $\emptyset$ and R^n are closed.*

14 **Problem**

Show that the interior of each set $\Gamma \subset R^n$ is an open set.

15 **Problem**

Show that the closure of each set $\Gamma \subset R^n$ is a closed set.

16 **Problem**

Let A be a fixed $m \times n$ matrix and b a fixed m-vector.

(i) Show that the set $\{x \mid x \in R^n, Ax \leq b\}$, and hence also the set $\{x \mid x \in R^n, Ax = b\}$ are closed sets in R^n.
(ii) Show that the set $\{x \mid x \in R^n, Ax < b\}$ is an open set in R^n.
(iii) Show that the set $\{x \mid x \in R^n, Ax \leq b\}$ is neither a closed nor an open set in R^n.

2. Sequences and bounds

1 **Sequence**

A *sequence* in a set X is a function f from the set I of all positive integers into the set X. If $f(n) = x^n \in X$ for $n \in I$, it is customary to denote the sequence f by the symbol $\{x^n\}$ or by $x^1, x^2, \ldots$. For any sequence of positive integers $n^1, n^2, \ldots$, such that $n^1 < n^2 < \cdots$, the sequence $x^{n^1}, x^{n^2}, \ldots$ is called a *subsequence* of $x^1, x^2, \ldots$. If $X = R^n$, then $x^1, x^2, \ldots$, is a sequence of points in R^n, and if $X = R$, then $x^1, x^2, \ldots$, is the familiar sequence of real numbers.

2   **Limit point**

Let x^1, x^2, . . . , be a sequence of points in R^n. A point $x \in R^n$ is said to be a *limit point* of the sequence if

$$\left\langle \begin{array}{l} \epsilon > 0 \\ \bar{n} = \text{positive integer} \end{array} \right\rangle \Rightarrow \|x^n - \bar{x}\| < \epsilon \text{ for some } n \geq \bar{n}$$

(In the literature, a limit point is sometimes called an *accumulation point* or a *cluster point*.)

3   **Limit**

Let x^1, x^2, . . . , be a sequence of points in R^n. A point $x^0 \in R^n$ is said to be a *limit* of the sequence, or we say that the sequence *converges* or *tends to a limit* x^0 if

$$\epsilon > 0 \Rightarrow \left\langle \begin{array}{l} \exists n^0: \\ \|x^n - x^0\| < \epsilon \text{ for all } n \geq n^0 \end{array} \right\rangle$$

We write

$$x^0 = \lim_{n \to \infty} x^n$$

4   **Remark**

Obviously a limit of a sequence is also a limit point of the sequence, but the converse is not necessarily true.   (For example the sequence 1, -1, 1, -1, . . . , has the limit points 1 and -1, but no limit.   The sequence 1, $\frac{1}{2}$, $\frac{1}{4}$, . . . , has a limit, and hence a limit point, zero.) Also note that if x^0 is a limit of x^1, x^2, . . . , then it is also a limit of each subsequence of x^1, x^2, ...

5   **Theorem**

(i)   *If $\bar{x}$ is a limit point of the sequence x^1, x^2, . . . , then there exists a subsequence x^{n^1}, x^{n^2}, . . . , which has $\bar{x}$ as a limit, and conversely.*

(ii)   *If a sequence x^1, x^2, . . . , tends to a limit x^0, then there can be no limit point (and hence no limit) other than x^0.*

PROOF   (i)   Let $\bar{x}$ be a limit point of x^1, x^2, ...   Then

$$\epsilon > 0 \Rightarrow \exists n^1: \|x^{n^1} - \bar{x}\| < \epsilon$$

$$\Rightarrow \exists n^2 \geq n^1 + 1: \|x^{n^2} - \bar{x}\| < \epsilon/2$$

and so on.   The subsequence x^{n^1}, x^{n^2}, . . . , converges to $\bar{x}$.

Conversely, if the subsequence x^{n^1}, x^{n^2}, . . . , converges to $\bar{x}$, then for given $\epsilon > 0$ and $\bar{n}$ there exists an $n^i \geq \bar{n}$ (in fact all $n^i \geq \bar{n}$) such that $\|x^{n^i} - \bar{x}\| < \epsilon$, and hence $\bar{x}$ is a limit point of x^1, x^2, ...

(ii) Let the sequence x^1, x^2, . . . , tend to the limit x^0, and $\bar{x} \neq x^0$. Then for $0 < \epsilon < \|\bar{x} - x^0\|$, there exists an integer n^0 such that

$$n \geq n^0 \Rightarrow \|x^n - x^0\| < \epsilon$$

$$\Rightarrow \epsilon + \|x^n - \bar{x}\| > \|x^0 - \bar{x}\| \qquad \text{(by triangle inequality, see } 1.3.10\text{)}$$

$$\Rightarrow \|x^n - \bar{x}\| > \|x^0 - \bar{x}\| - \epsilon > 0$$

and hence $\bar{x}$ is not a limit point of the sequence. ■

6   **Remark**

Note that a limit point (and hence a limit) of a sequence x^1, x^2, . . . , in R^n is a point of closure of any set Γ in R^n containing $\{x^1, x^2, \ldots\}$. Conversely, if $\bar{x}$ is a point of closure of the set $\Gamma \subset R^n$, then there exists a sequence x^1, x^2, . . . $\subset \Gamma$ (and hence also a subsequence x^{n^1}, x^{n^2}, . . .) such that $\bar{x}$ is a limit point of x^1, x^2, . . . (and hence a limit of x^{n^1}, x^{n^2}, . . .).

7   **Lower and upper bounds**

Let Γ be a nonempty set of real numbers. Then

α is a *lower bound* of $\Gamma \Leftrightarrow \langle x \in \Gamma \Rightarrow x \geq \alpha \rangle$

β is an *upper bound* of $\Gamma \Leftrightarrow \langle x \in \Gamma \Rightarrow x \leq \beta \rangle$

8   **Greatest lower and least upper bounds**

Let Γ be a nonempty set of real numbers. A lower bound $\bar{\alpha}$ is a *greatest lower bound* of Γ (*infimum* of Γ, or inf Γ) if no bigger number is a lower bound of Γ. An upper bound β is a *least upper bound* of Γ (*supremum* of Γ, or sup Γ) if no smaller number is an upper bound of Γ. Equivalently we have

$$\bar{\alpha} = \inf \Gamma \Leftrightarrow \left\langle \begin{array}{l} x \in \Gamma \Rightarrow x \geq \bar{\alpha} \\ \epsilon > 0 \Rightarrow \exists x \in \Gamma : \bar{\alpha} + \epsilon > x \end{array} \right\rangle$$

$$\bar{\beta} = \sup \Gamma \Leftrightarrow \left\langle \begin{array}{l} x \in \Gamma \Rightarrow x \leq \bar{\beta} \\ \epsilon > 0 \Rightarrow \exists x \in \Gamma : \bar{\beta} - \epsilon < x \end{array} \right\rangle$$

We shall take the following axiom as one of the axioms of the real number system [Birkhoff-Maclane 53].

9   **Axiom**

Any nonempty set Γ of real numbers which has a lower (upper) bound has a greatest (least) lower (upper) bound.

If the set Γ has no infimum (or equivalently by the above axiom if it has no lower bound), we say that Γ is *unbounded from below* and we write inf $\Gamma = -\infty$. Similarly if Γ has no supremum (or equivalently by the above axiom if it has no upper bound), we say Γ is *unbounded from above* and we write sup $\Gamma = +\infty$. Hence by augmenting the Euclidean line R by the two points $+\infty$ and $-\infty$ any nonempty set Γ will have an infimum, which may be $-\infty$, and a supremum, which may be $+\infty$. We shall follow the convention of writing inf $\emptyset = +\infty$ and sup $\emptyset = -\infty$.

We observe that neither inf Γ nor sup Γ need be in Γ. For example inf $\{1, \frac{1}{2}, \frac{1}{3}, \ldots\}$ is 0, but 0 is not in the set $\{1, \frac{1}{2}, \frac{1}{3}, \ldots\}$.

10   **Theorem**

Every bounded nondecreasing (nonincreasing) sequence of real numbers has a limit.

PROOF Let $x^1, x^2, \ldots$, be a bounded nondecreasing sequence of real numbers. By the above Axiom *9*, the sequence has a least upper bound $\bar{\beta}$. Hence

$$\epsilon > 0 \Rightarrow \exists \bar{n}: \bar{\beta} - x^{\bar{n}} < \epsilon \qquad \text{(by 8)}$$

$$\Rightarrow \bar{\beta} - x^n < \epsilon \text{ for } n \geq \bar{n} \qquad \text{(because sequence is non-decreasing)}$$

$$\Rightarrow -\epsilon < 0 \leq \bar{\beta} - x^n < \epsilon \text{ for } n \geq \bar{n} \qquad \text{(because } \bar{\beta} \geq x^n \text{ for all } n\text{)}$$

$$\Rightarrow \bar{\beta} \text{ is a limit of } x^1, x^2, \ldots$$

The proof for a bounded nonincreasing sequence is similar. ∎

11   **Cauchy convergence criterion**

A sequence $x^1, x^2, \ldots$, in R^n converges to a limit x^0 if and only if it is a Cauchy sequence, that is, for each $\epsilon > 0$ there exists an n^* such that $\|x^m - x^n\| < \epsilon$ for each $m,n \geq n^*$.

For a proof see [Buck 65, Fleming 65, Rudin 64].

3. Compact sets in R^n

1   **Bounded set**

A set $\Gamma \subset R^n$ is *bounded* if there exists a real number α such that for each $x \in \Gamma$, $\|x\| \leq \alpha$.

A set $\Gamma \subset R^n$ which is both closed and bounded has some very interesting properties. In fact such sets constitute one of the most important classes of sets in analysis, the class of *compact sets*. We give in the theorem below a number of equivalent characterizations of a compact set in R^n. (These characterizations are not necessarily equivalent in spaces more general than R^n.)

2 Compact sets (theorem-definition)

A set $\Gamma \subset R^n$ is said to be compact if it satisfies any of the following equivalent conditions:

(i) Γ *is closed and bounded.*

(ii) *(Bolzano-Weierstrass) Every sequence of points in Γ has a limit point in Γ.*

(iii) *(Finite intersection property) For any family $(\Gamma_i)_{i \in I}$ of sets, closed relative to Γ, it follows that*

$$\bigcap_{i \in I} \Gamma_i = \emptyset \Rightarrow \left\langle \begin{array}{l} \Gamma_{i_1} \cap \Gamma_{i_2} \cdots \cap \Gamma_{i_m} = \emptyset \\ \textit{for some } i_1, i_2, \ldots, i_m \in I \end{array} \right\rangle$$

or equivalently

$$\bigcap_{i \in I} \Gamma_i \neq \emptyset \Leftarrow \left\langle \begin{array}{l} \Gamma_{i_1} \cap \Gamma_{i_2} \cdots \cap \Gamma_{i_m} \neq \emptyset \\ \textit{for all } i_1, i_2, \ldots, i_m \in I \end{array} \right\rangle$$

(iv) *(Heine-Borel)   From every family $(\Gamma_i)_{i \in I}$ of open sets whose union $\bigcup_{i \in I} \Gamma_i$ contains Γ we can extract a finite subfamily $\Gamma_{i_1}, \ldots, \Gamma_{im}$ whose union $\Gamma_{i_1} \cup \Gamma_{i_2} \cdots \cup \Gamma_{i_m}$ contains Γ (or equivalently stated: every open covering of Γ has a finite subcovering).*

We shall not give here a proof of the equivalence of the above four conditions; such a proof can be found in any of the references given at the beginning of this Appendix. An especially lucid proof is also given in chap. 2 of [Berge–Ghouila Houri 65].

3 Corollary

Let Γ and Λ be sets in R^n which are respectively compact and closed. Then the sum Ω

$$\Omega = \Gamma + \Lambda = \{x + y \mid x \in \Gamma, y \in \Lambda\}$$

is closed.

PROOF   Let x^0 belong to the closure of Ω. Then there exists a sequence $x^1, x^2, \ldots,$ in Ω which converges to x^0 (see *B.2.6*). Then we can find a sequence $y^1, y^2, \ldots$, in Γ and a sequence $z^1, z^2, \ldots$, in Λ such that

$$x^n = y^n + z^n \quad \text{for } n = 1, 2, \ldots$$

Since Γ is compact, there exists a subsequence $y^{n^1}, y^{n^2}, \ldots$, which converges to $y^0 \in \Gamma$. Then

$$\lim_{i \to \infty} x^{n^i} = x^0 \qquad \text{(see } B.2.4\text{)}$$

$$\lim_{i \to \infty} y^{n^i} = y^0 \in \Gamma$$

$$\lim_{i \to \infty} z^{n^i} = x^0 - y^0 \in \Lambda \qquad \text{(Since } \Lambda \text{ is closed, see } B.2.6\text{)}$$

Hence

$$x^0 = y^0 + (x^0 - y^0) \in \Omega$$

and Ω is closed. ∎

Appendix C

Continuous and Semicontinuous Functions, Minima and Infima

We collect in this appendix some basic definitions and properties of continuous and semicontinuous functions defined on a subset of R^n. We also state some facts about the minima and maxima of such functions. More detailed discussions can be found in [Berge 63, Berge–Ghouila Houri 65.]

1. Continuous and semi-continuous functions

Continuous function

A numerical function θ defined on a set $\Gamma \subset R^n$ is said to be *continuous at* $x^0 \in \Gamma$ (with respect to Γ) if either of the two following equivalent conditions are satisfied:

(i) For each $\epsilon > 0$ there exists a $\delta > 0$ such that

$$\left.\begin{array}{c} \|x - x^0\| < \delta \\ x \in \Gamma \end{array}\right\} \Rightarrow$$
$$-\epsilon < \theta(x) - \theta(x^0) < \epsilon$$

(ii) For each sequence x^1, x^2, . . . , in Γ converging to x^0,
$$\lim_{n \to \infty} \theta(x^n) = \theta(\lim_{n \to \infty} x^n) = \theta(x^0)$$

θ is said to be *continuous on* Γ (with respect to Γ) if it is continuous (with respect to Γ) at each point $x^0 \in \Gamma$, or equivalently if any of the following equivalent conditions hold:

(iii) The sets
$$\{x \mid x \in \Gamma, \theta(x) \leqq \alpha\} \quad \text{and}$$
$$\{x \mid x \in \Gamma, \theta(x) \geqq \alpha\}$$

are closed relative to Γ for each real α.

(iv) The sets

$$\{x \mid x \in \Gamma, \; \theta(x) > \alpha\} \qquad \text{and} \qquad \{x \mid x \in \Gamma, \; \theta(x) < \alpha\}$$

are open relative to Γ for each real α.

(v)   The epigraph of θ

$$G_\theta = \{(x,\varsigma) \mid x \in \Gamma, \; \varsigma \in R, \; \theta(x) \leq \varsigma\}$$

and the hypograph of θ

$$H_\theta = \{(x,\varsigma) \mid x \in \Gamma, \; \varsigma \in R, \; \theta(x) \geq \varsigma\}$$

are closed relative to $\Gamma \times R$.

In each of the above conditions a pair of symmetric requirements can be found.   By dropping one requirement from each pair, we arrive at the concept of semicontinuous functions.

2   **Lower semicontinuous function**

A numerical function θ defined on a set $\Gamma \subset R^n$ is said to be *lower semicontinuous at $x^0 \in \Gamma$* (with respect to Γ) if either of the two following equivalent conditions is satisfied:

(i)   For each $\epsilon > 0$ there exists a $\delta > 0$ such that

$$\left. \begin{array}{c} \|x - x^0\| < \delta \\[2mm] x \in \Gamma \end{array} \right\} \Rightarrow -\epsilon < \theta(x) - \theta(x^0)$$

(ii)   For each sequence $x^1, x^2, \ldots$, in Γ converging to x^0,

$$\liminf_{n \to \infty} \theta(x^n) \geq \theta(\lim_{n \to \infty} x^n) = \theta(x^0)$$

where $\liminf\limits_{n \to \infty} \theta(x^n)$ denotes the infimum of the limit points of the sequence of real numbers $\theta(x^1), \theta(x^2), \ldots$.

θ is said to be *lower semicontinuous on Γ* (with respect to Γ) if it is lower semicontinuous (with respect to Γ) at each point $x^0 \in \Gamma$, or equivalently if any of the following equivalent conditions holds:

(iii)   The set

$$\{x \mid x \in \Gamma, \; \theta(x) \leq \alpha\}$$

is closed relative to Γ for each real α.

(iv)   The set

$$\{x \mid x \in \Gamma,\ \theta(x) > \alpha\}$$

is open relative to Γ for each real α.

(v)   The epigraph of θ

$$G_\theta = \{(x,\varsigma) \mid x \in \Gamma,\ \varsigma \in R,\ \theta(x) \leq \varsigma\}$$

is closed relative to $\Gamma \times R$.

3

Upper semicontinuous function

A numerical function θ defined on a set $\Gamma \subset R^n$ is said to be *upper semicontinuous at* $x^0 \in \Gamma$ (with respect to Γ) if either of the two following equivalent conditions is satisfied:

(i)   For each $\epsilon > 0$ there exists a $\delta > 0$ such that

$$\left\langle \begin{array}{c} \|x - x^0\| < \delta \\ x \in \Gamma \end{array} \right\rangle \Rightarrow \theta(x) - \theta(x^0) < \epsilon$$

(ii)   For each sequence $x^1,\ x^2,\ \ldots\ ,$ in Γ converging to x^0,

$$\limsup_{n \to \infty} \theta(x^n) \leq \theta(\lim_{n \to \infty} x^n) = \theta(x^0)$$

where $\limsup\limits_{n \to \infty} \theta(x^n)$ denotes the supremum of the limit points of the sequence of real numbers $\theta(x^1),\ \theta(x^2),\ \ldots\ .$

θ is said to be *upper semicontinuous on* Γ (with respect to Γ) if it is upper semicontinuous (with respect to Γ) at each point $x^0 \in \Gamma$, or equivalently if any of the following equivalent conditions hold:

(iii) The set

$$\{x \mid x \in \Gamma,\ \theta(x) \geq \alpha\}$$

is closed relative to Γ for each real α.

(iv)   The set

$$\{x \mid x \in \Gamma,\ \theta(x) < \alpha\}$$

is open relative to Γ for each real α.

(v)   The hypograph of θ

$$H_\theta = \{(x,\varsigma) \mid x \in \Gamma,\ \varsigma \in R,\ \theta(x) \geq \varsigma\}$$

is closed relative to $\Gamma \times R$.

4

Remark

θ is lower semicontinuous at $x^0 \in \Gamma$ (with respect to Γ) if and only if $-\theta$ is upper semicontinuous at $x^0 \in \Gamma$ (with respect to Γ). θ is continuous at $x^0 \in \Gamma$ (with respect to Γ) if and only if it is both lower semicontinuous and upper semicontinuous at $x^0 \in \Gamma$ (with respect to Γ).

5   **Examples**

(i) $\theta(x) = \left\langle \begin{array}{l} x, \text{ for } x \neq 1 \\ \tfrac{1}{2}, \text{ for } x = 1 \end{array} \right\rangle$ is lower semicontinuous on R, see Fig. C.1.1

(ii) $\theta(x) = \left\langle \begin{array}{l} (x)^2, \text{ for } x \neq 0 \\ \tfrac{1}{2}, \text{ for } x = 0 \end{array} \right\rangle$ is upper semicontinuous on R, see Fig. C.1.2

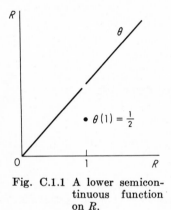

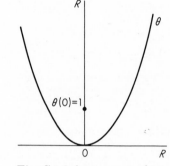

Fig. C.1.1 A lower semicontinuous function on R.

Fig. C.1.2 An upper semicontinuous function on R.

6   **Theorem**

Let $(\theta_i)_{i \in I}$ be a (finite or infinite) family of lower semicontinuous functions on $\Gamma \subset R^n$. Its least upper bound

$$\theta(x) = \sup_{i \in I} \theta_i(x)$$

is lower semicontinuous on Γ. If I is finite, then its greatest lower bound

$$\phi(x) = \inf_{i \in I} \theta_i(x)$$

is also lower semicontinuous on Γ.

PROOF The first part follows directly from 2(iii) above and Theorem B.1.13(i) if we observe that for any real λ, the set

$$\{x \mid \theta(x) \leq \lambda\} = \{x \mid \sup_{i \in I} \theta_i(x) \leq \lambda\} = \bigcap_{i \in I} \{x \mid \theta_i(x) \leq \lambda\}$$

is closed (relative to Γ).   The second part follows from 2(iii) and Theorem $B.1.13$(ii) if we observe that for any real λ the set

$$\{x \mid \phi(x) \leqq \lambda\} \;=\; \{x \mid \inf_{i \in I} \theta_i(x) \leqq \lambda\} \;=\; \bigcup_{i \in I} \{x \mid \theta_i(x) \leqq \lambda\}$$

is closed (relative to Γ).   ∎

7   Corollary

Let $(\theta_i)_{i \in I}$ be a (finite or infinite) family of upper semicontinuous functions on $\Gamma \subset R^n$.   Its greatest lower bound

$$\phi(x) \;=\; \inf_{i \in I} \theta_i(x)$$

is upper semicontinuous on Γ.   If I is finite, then its least upper bound

$$\theta(x) \;=\; \sup_{i \in I} \theta_i(x)$$

is also upper semicontinuous on Γ.

2.   Infimum (supremum) and minimum (maximum) of a set of real numbers

We recall that in Appendix B $(B.2.8)$ we defined the infimum and supremum of a set of real numbers Γ as follows

$$1 \qquad \bar{\alpha} = \inf \Gamma \Leftrightarrow \left\langle \begin{array}{l} x \in \Gamma \Rightarrow x \geqq \bar{\alpha} \\[2mm] \epsilon > 0 \Rightarrow \exists x \in \Gamma : \bar{\alpha} + \epsilon > x \end{array} \right\rangle$$

and

$$2 \qquad \bar{\beta} = \sup \Gamma \Leftrightarrow \left\langle \begin{array}{l} x \in \Gamma \Rightarrow x \leqq \bar{\beta} \\[2mm] \epsilon > 0 \Rightarrow \exists x \in \Gamma : \bar{\beta} - \epsilon < x \end{array} \right\rangle$$

We also noted that neither $\bar{\alpha}$ nor $\bar{\beta}$ need be in Γ.   However when $\bar{\alpha}$ and $\bar{\beta}$ are in Γ, they are called respectively min Γ and max Γ.

3   Minimum (maximum)

Let Γ be a set of real numbers.   If (inf Γ) $\in \Gamma$, then inf Γ is called the *minimum of* Γ and is denoted by min Γ.   If (sup Γ) $\in \Gamma$, then sup Γ is called the *maximum of* Γ and is denoted by max Γ.   Equivalently

$$\bar{\alpha} = \min \Gamma \Leftrightarrow \left\langle \begin{array}{l} \bar{\alpha} \in \Gamma \\[2mm] x \in \Gamma \Rightarrow x \geqq \bar{\alpha} \end{array} \right\rangle$$

$$\bar{\beta} = \max \Gamma \Leftrightarrow \left\langle \begin{array}{l} \bar{\beta} \in \Gamma \\[2mm] x \in \Gamma \Rightarrow x \leqq \bar{\beta} \end{array} \right\rangle$$

3. Infimum (supremum) and minimum (maximum) of a numerical function

1   **Bounded functions**

A numerical function θ defined on the set Γ is said to be *bounded from below on* Γ if there exists a number α such that

$$x \in \Gamma \Rightarrow \theta(x) \geqq \alpha$$

The number α is a *lower bound of* θ on Γ. θ is said to be *bounded from above on* Γ if there exists a number β such that

$$x \in \Gamma \Rightarrow \theta(x) \leqq \beta$$

The number β is an *upper bound of* θ on Γ.

2   **Infimum of a numerical function**

Let θ be a numerical function defined on the set Γ.   If there is a number $\bar{\alpha}$ such that

$$x \in \Gamma \Rightarrow \theta(x) \geqq \bar{\alpha}$$

and

$$\epsilon > 0 \Rightarrow \langle \exists x \in \Gamma \colon \bar{\alpha} + \epsilon > \theta(x) \rangle$$

then $\bar{\alpha}$ is called the *infimum of θ on* Γ, and we write

$$\bar{\alpha} = \inf_{x \in \Gamma} \theta(x)$$

3   **Supremum of a numerical function**

Let θ be a numerical function defined on the set Γ.   If there is a number $\bar{\beta}$ such that

$$x \in \Gamma \Rightarrow \theta(x) \leqq \bar{\beta}$$

and

$$\epsilon > 0 \Rightarrow \langle \exists x \in \Gamma \colon \bar{\beta} - \epsilon < \theta(x) \rangle$$

then $\bar{\beta}$ is called the *supremum of θ on* Γ, and we write

$$\bar{\beta} = \sup_{x \in \Gamma} \theta(x)$$

If we admit the points $\pm \infty$, then every numerical function θ has a supremum and infimum on the set Γ on which it is defined.

4   Examples

$$\inf_{x \in R} e^{-x} = 0 \qquad \sup_{x \in R} e^{-x} = +\infty$$

$$\inf_{x \in R} x = -\infty \qquad \sup_{x \in R} x = +\infty$$

$$\inf_{x \in R} \sin x = -1 \qquad \sup_{x \in R} \sin x = 1$$

5   Minimum of a numerical function

Let θ be a numerical function defined on the set Γ.   If there exists an $\bar{x} \in \Gamma$ such that

$$x \in \Gamma \Rightarrow \theta(x) \geq \theta(\bar{x})$$

then $\theta(\bar{x})$ is called the *minimum of θ on* Γ, and we write

$$\theta(\bar{x}) = \min_{x \in \Gamma} \theta(x)$$

We make the following remarks:

(i)   Not every numerical function has a minimum; for example, e^{-x} and x have no minima on R.

(ii)   The minimum of a numerical function, if it exists, must be finite.

(iii)   The minimum of a numerical function, if it exists, is an attained infimum, that is,

$$\theta(\bar{x}) = \min_{x \in \Gamma} \theta(x) = \inf_{x \in \Gamma} \theta(x)$$

6   Maximum of a numerical function

Let θ be a numerical function defined on the set Γ.   If there exists an $\bar{x} \in \Gamma$ such that

$$x \in \Gamma \Rightarrow \theta(x) \leq \theta(\bar{x})$$

then $\theta(\bar{x})$ is called the *maximum of θ on* Γ and we write

$$\theta(\bar{x}) = \max_{x \in \Gamma} \theta(x)$$

Remarks similar to (i), (ii), and (iii) above, which applied to the minimum of a function, also apply here to the maximum of a function.

4. Existence of a minimum and a maximum of a numerical function

We give now sufficient conditions for a numerical function defined on a subset of R^n to have a minimum and a maximum on that subset.

1

Theorem

   A lower (upper) semicontinuous function θ defined on a compact— that is, closed and bounded—set Γ in R^n is bounded from below (above) and attains in Γ the value

$$\bar{\alpha} = \inf_{x \in \Gamma} \theta(x) \qquad [\bar{\beta} = \sup_{x \in \Gamma} \theta(x)]$$

In other words there exists an $\bar{x} \in \Gamma$ such that

$$x \in \Gamma \Rightarrow \theta(x) \geqq \theta(\bar{x}) \qquad [x \in \Gamma \Rightarrow \theta(x) \leqq \theta(\bar{x})]$$

PROOF We prove the lower semicontinuous case. Let $\gamma > \bar{\alpha}$. Then the set

$$\Lambda_\gamma = \{x \mid x \in \Gamma, \, \theta(x) \leqq \gamma\}$$

is not empty and is closed relative to Γ by *C.1.2*(iii). Hence by the finite intersection theorem *B.3.2*(iii)

$$\bigcap_{\gamma > \alpha} \Lambda_\gamma \neq \emptyset$$

Choose $\bar{x}$ in this intersection; then $\theta(\bar{x}) = \bar{\alpha}$, and hence $\bar{\alpha}$ is finite. ∎
   It should be remarked here that the above theorem cannot be strengthened by dropping the semicontinuity or the compactness assumptions. In other words, in order to ensure that

$$\inf_{x \in \Gamma} \theta(x) = \min_{x \in \Gamma} \theta(x)$$

we cannot weaken the conditions that

   (i) θ is lower semicontinuous on Γ,
  (ii) Γ is closed, and
 (iii) Γ is bounded.

We give examples below where the infimum is not attained whenever any one of the above conditions is violated.

(ī) $\theta(x) = \begin{cases} \frac{1}{2} & \text{for } x = 0,\, x \in R \\ x & \text{for } 0 < x \leq 1,\, x \in R \end{cases}$

$\inf\limits_{0 \leq x \leq 1} \theta(x) = 0$, but no minimum exists on the compact set $\{x \mid 0 \leq x \leq 1\}$.

(īī) $\theta(x) = x$ for $0 < x < 1,\, x \in R$

$\inf\limits_{0 < x < 1} \theta(x) = 0$, but no minimum exists of this continuous function on the open set $\{x \mid 0 < x < 1\}$.

(īīī) $\theta(x) = e^{-x}$ $x \in R$

$\inf\limits_{x \in E} \theta(x) = 0$, but no minimum exists of this continuous function on the closed unbounded set R.

Appendix D

Differentiable Functions, Mean-value and Implicit Function Theorems

We summarize here the pertinent definitions and properties of differentiable functions and give the mean-value and implicit function theorems. For details the reader is referred to [Fleming 65, Rudin 64, Hestenes 66, Bartle 64].

1. Differentiable and twice-differentiable functions

1   **Differentiable numerical function**

Let θ be a numerical function defined on an open set Γ in R^n, and let $\bar{x}$ be in Γ. θ is said to be *differentiable at* $\bar{x}$ if for all $x \in R^n$ such that $\bar{x} + x \in \Gamma$ we have that

$$\theta(\bar{x} + x) = \theta(\bar{x}) + t(\bar{x})x + \alpha(\bar{x},x)\|x\|$$

where $t(\bar{x})$ is an n-dimensional bounded vector, and α is a numerical function of x such that

$$\lim_{x \to 0} \alpha(\bar{x},x) = 0$$

θ is said to be *differentiable on* Γ if it is differentiable at each $\bar{x}$ in Γ. [Obviously if θ is differentiable on the open set Γ, it is also differentiable on any subset Λ (open or not) of Γ. Hence when we say that θ is differentiable on some set Λ (open or not), we shall mean that θ is differentiable on some open set containing Λ.]

2   **Partial derivatives and gradient of a numerical function**

Let θ be a numerical function defined on an open set Γ in R^n, and let $\bar{x}$ be in Γ. θ is said to have a *partial derivative at* $\bar{x}$ with respect to x_i, $i = 1, \ldots, n$, if

$$\frac{\theta(\bar{x}_1, \ldots, \bar{x}_{i-1}, \bar{x}_i + \delta, \bar{x}_{i+1}, \ldots, \bar{x}_n) - \theta(\bar{x}_1, \ldots, \bar{x}_n)}{\delta}$$

tends to a finite limit when δ tends to zero. This limit is called the *partial derivative of θ with respect to x_i at $\bar{x}$* and is denoted by $\partial\theta(\bar{x})/\partial x_i$. The n-dimensional vector of the partial derivatives of θ with respect to $x_1, \ldots, x_n$ at $\bar{x}$ is called the *gradient of θ at $\bar{x}$* and is denoted by $\nabla\theta(\bar{x})$, that is,

$$\nabla\theta(\bar{x}) = \left(\frac{\partial\theta(\bar{x})}{\partial x_1}, \ldots, \frac{\partial\theta(\bar{x})}{\partial x_n}\right)$$

3   **Theorem**

Let θ be a numerical function defined on an open set Γ in R^n, and let $\bar{x}$ be in Γ.

(i)   *If θ is differentiable at $\bar{x}$, then θ is continuous at $\bar{x}$, and $\nabla\theta(\bar{x})$ exists (but not conversely), and*

$$\left.\begin{array}{c} \theta(\bar{x} + x) = \theta(\bar{x}) + \nabla\theta(\bar{x})x + \alpha(\bar{x},x)\|x\| \\ \lim_{x\to 0} \alpha(\bar{x},x) = 0 \end{array}\right\} \quad for\ \bar{x} + x \in \Gamma$$

(ii)   *If θ has continuous partial derivatives at $\bar{x}$ with respect to $x_1, \ldots, x_n$, that is, $\nabla\theta(\bar{x})$ exists and $\nabla\theta$ is continuous at $\bar{x}$, then θ is differentiable at $\bar{x}$.*

We can summarize most of the above result as follows

$$\langle\theta\ \text{differentiable at}\ \bar{x}\rangle \genfrac{}{}{0pt}{}{\Rightarrow}{\not\Leftarrow} \left\{\begin{array}{l} \theta\ \text{continuous at}\ \bar{x} \\ \nabla\theta(\bar{x})\ \text{exists} \end{array}\right.$$

$$\langle\theta\ \text{differentiable at}\ \bar{x}\rangle \Leftarrow \left\{\begin{array}{l} \nabla\theta(\bar{x})\ \text{exists} \\ \nabla\theta\ \text{continuous at}\ \bar{x} \end{array}\right.$$

4   **Differentiable vector function**

Let f be an m-dimensional vector function defined on an open set Γ in R^n, and let $\bar{x}$ be in Γ. f is said to be *differentiable at $\bar{x}$* (respectively *on Γ*) if each of its components $f_1, \ldots, f_m$ is differentiable at $\bar{x}$ (respectively on Γ).

5   **Partial derivatives and Jacobian of a vector function**

Let f be an m-dimensional vector function defined on an open set Γ in R^n, and let $\bar{x}$ be in Γ. f is said to have partial derivatives at $\bar{x}$ with respect to $x_1, \ldots, x_n$ if each of its components $f_1, \ldots, f_m$ has partial derivatives at $\bar{x}$ with respect to $x_1, \ldots, x_n$. We write

$$\nabla f(\bar{x}) = \begin{bmatrix} \dfrac{\partial f_1(\bar{x})}{\partial x_1} & \cdots & \dfrac{\partial f_1(\bar{x})}{\partial x_n} \\ \cdot & & \cdot \\ \cdot & & \cdot \\ \cdot & & \cdot \\ \dfrac{\partial f_m(\bar{x})}{\partial x_1} & \cdots & \dfrac{\partial f_m(\bar{x})}{\partial x_n} \end{bmatrix}$$

The $m \times n$ matrix $\nabla f(\bar{x})$ is called the *Jacobian (matrix) of f at $\bar{x}$.*

6   **Chain rule theorem**

Let f be an m-dimensional vector function defined on an open set Γ in R^n, and let ϕ be a numerical function which is defined on R^m. The numerical function θ defined on Γ by

$$\theta(x) = \phi[f(x)]$$

is differentiable at $\bar{x} \in \Gamma$ if f is differentiable at $\bar{x}$ and if ϕ is differentiable at $\bar{y} = f(\bar{x})$, and

$$\nabla\theta(\bar{x}) = \nabla\phi(\bar{y})\nabla f(\bar{x})$$

7   **Twice-differentiable numerical function and its Hessian**

Let θ be a numerical function defined on an open set Γ in R^n, and let $\bar{x}$ be in Γ. θ is said to be twice differentiable at $\bar{x}$ if for all $x \in R^n$ such that $\bar{x} + x \in \Gamma$ we have that

$$\theta(\bar{x} + x) = \theta(\bar{x}) + \nabla\theta(\bar{x})x + \frac{x\nabla^2\theta(\bar{x})x}{2} + \beta(\bar{x},x)(\|x\|)^2$$

where $\nabla^2\theta(\bar{x})$ is an $n \times n$ matrix of bounded elements, and β is a numerical function of x such that

$$\lim_{x \to 0} \beta(\bar{x},x) = 0$$

The $n \times n$ matrix $\nabla^2\theta(\bar{x})$ is called the Hessian (matrix) of θ at $\bar{x}$ and its ijth element is written as

$$[\nabla^2\theta(\bar{x})]_{ij} = \frac{\partial^2\theta(\bar{x})}{\partial x_i \, \partial x_j} \qquad i,j = 1, \ldots, n$$

Obviously if θ is twice differentiable at $\bar{x}$, it must also be differentiable at $\bar{x}$.

8   **Theorem**

Let θ be a numerical function defined on an open set Γ in R^n, and let $\bar{x}$ be in Γ. Then

(i) $\nabla\theta$ *differentiable at* $\bar{x} \Rightarrow \theta$ *twice differentiable at* $\bar{x}$

(ii) $\nabla\theta$ *has continuous partial derivatives at* $\bar{x} \Rightarrow \theta$ *twice differentiable at* $\bar{x}$

(iii) $\left\langle \nabla^2\theta \text{ continuous at } \bar{x} \right\rangle \Rightarrow \left\langle \begin{array}{l} \dfrac{\partial^2\theta(\bar{x})}{\partial x_i\, \partial x_j} = \dfrac{\partial}{\partial x_i}\left(\dfrac{\partial\theta(\bar{x})}{\partial x_j}\right) \\[2mm] \text{and } \nabla^2\theta(\bar{x}) \text{ is symmetric,} \\[1mm] \text{that is, } [\nabla^2\theta(\bar{x})]_{ij} = [\nabla^2\theta(\bar{x})]_{ji} \end{array} \right\rangle$

9   Remark

The numbers $\partial\theta(\bar{x})/\partial x_i$, $i = 1, \ldots, n$, are also called the first partial derivatives of θ at $\bar{x}$, and $\partial^2\theta(\bar{x})/\partial x_i\, \partial x_j\ i,j = 1, \ldots, n$, are also called the second partial derivatives of θ at $\bar{x}$.   In an analogous way we can define kth partial derivatives of θ at $\bar{x}$.

10   Remark

Let θ be a numerical function defined on an open set $\Gamma \subset R^n \times R^k$ which is differentiable at $(\bar{x},\bar{y}) \in \Gamma$.   We define then

$$\nabla_x\theta(\bar{x},\bar{y}) = \left[\frac{\partial\theta(\bar{x},\bar{y})}{\partial x_1}, \ldots, \frac{\partial\theta(\bar{x},\bar{y})}{\partial x_n}\right]$$

$$\nabla_y\theta(\bar{x},\bar{y}) = \left[\frac{\partial\theta(\bar{x},\bar{y})}{\partial y_1}, \ldots, \frac{\partial\theta(\bar{x},\bar{y})}{\partial y_k}\right]$$

and

$$\nabla\theta(\bar{x},\bar{y}) = [\nabla_x\theta(\bar{x},\bar{y}),\, \nabla_y\theta(\bar{x},\bar{y})]$$

Let f be an m-dimensional function defined on an open set $\Gamma \subset R^n \times R^k$ which is differentiable at $(\bar{x},\bar{y}) \in \Gamma$.   We define then

$$\nabla_x f(\bar{x},\bar{y}) = \begin{bmatrix} \dfrac{\partial f_1(\bar{x},\bar{y})}{\partial x_1} & \cdots & \dfrac{\partial f_1(\bar{x},\bar{y})}{\partial x_n} \\ \cdot & & \cdot \\ \cdot & & \cdot \\ \cdot & & \cdot \\ \dfrac{\partial f_m(\bar{x},\bar{y})}{\partial x_1} & \cdots & \dfrac{\partial f_m(\bar{x},\bar{y})}{\partial x_n} \end{bmatrix}$$

$$\nabla_y f(\bar{x},\bar{y}) = \begin{bmatrix} \dfrac{\partial f_1(\bar{x},\bar{y})}{\partial y_1} & \cdots & \dfrac{\partial f_1(\bar{x},\bar{y})}{\partial y_k} \\ \cdot & & \cdot \\ \cdot & & \cdot \\ \cdot & & \cdot \\ \dfrac{\partial f_m(\bar{x},\bar{y})}{\partial y_1} & \cdots & \dfrac{\partial f_m(\bar{x},\bar{y})}{\partial y_k} \end{bmatrix}$$

and

$$\nabla f(\bar{x},\bar{y}) = [\nabla_x f(\bar{x},\bar{y})\quad \nabla_y f(\bar{x},\bar{y})]$$

2.   Mean-value theorem and Taylor's theorem

1   Mean-value theorem

Let θ be a differentiable numerical function on the open convex set $\Gamma \subset R^n$, and let $x^1, x^2 \in \Gamma$.   Then

$$\theta(x^2) = \theta(x^1) + \nabla\theta[x^1 + \gamma(x^2 - x^1)](x^2 - x^1)$$

for some real number γ, $0 < \gamma < 1$.

2   Taylor's theorem (second order)

Let θ be a twice-differentiable numerical function on the open convex set $\Gamma \subset R^n$, and let $x^1, x^2 \in \Gamma$.   Then

$$\theta(x^2) = \theta(x^1) + \nabla\theta(x^1)(x^2 - x^1) + \frac{(x^2 - x^1)\nabla^2\theta[x^1 + \delta(x^2 - x^1)](x^2 - x^1)}{2}$$

for some real number δ, $0 < \delta < 1$.

3.   Implicit function theorem

1   Implicit function theorem

Let f be an m-dimensional vector function defined on an open set $\Lambda \subset R^n \times R^m$, let f have continuous first partial derivatives at $(\bar{x}, \bar{y}) \in \Lambda$, let $f(\bar{x}, \bar{y}) = 0$, and let $\nabla_y f(\bar{x}, \bar{y})$ be nonsingular.

Then there exist a ball $B_\lambda(\bar{x}, \bar{y})$ with radius $\lambda > 0$ in R^{n+m}, an open set $\Gamma \subset R^n$ containing $\bar{x}$, and an m-dimensional vector function e with continuous first partial derivatives on Γ such that

$\nabla_y f(x, y)$ is nonsingular for all $(x, y) \in B_\lambda(\bar{x}, \bar{y})$

$\bar{y} = e(\bar{x})$

and

$f[x, e(x)] = 0$ for $x \in \Gamma$

References

Abadie, J.: On the Kuhn-Tucker Theorem, in J. Abadie (ed.), "Nonlinear Programming," pp. 21–36, North Holland Publishing Company, Amsterdam, 1967.

Anderson, K. W., and D. W. Hall: "Sets, Sequences and Mappings, the Basic Concepts of Analysis," John Wiley & Sons, Inc., New York, 1963.

Arrow, K. J., and A. C. Enthoven: Quasiconcave Programming, *Econometrica* **29**: 779–800 (1961).

Arrow, K. J., L. Hurwicz, and H. Uzawa (eds.): "Studies in Linear and Nonlinear Programming," Stanford University Press, Stanford, Calif., 1958.

Arrow, K. J., L. Hurwicz, and H. Uzawa: Constraint Qualifications in Maximization Problems, *Naval Research Logistics Quarterly* **8**: 175–191 (1961).

Bartle, R. G.: "The Elements of Real Analysis," John Wiley & Sons, Inc., New York, 1964.

Berge, C.: "Topological Spaces," The MacMillan Company, New York, 1963.

Berge, C., and A. Ghouila Houri: "Programming, Games and Transportation Networks," John Wiley & Sons, Inc., New York, 1965.

Birkhoff, G. and S. Maclane: "A Survey of Modern Algebra," The MacMillan Company, New York, 1953.

Bohnenblust, H. F., S. Karlin, and L. S. Shapley: "Solutions of Discrete, Two-person Games," Contributions to the Theory of Games, vol. I, Annals of Mathematics Studies Number 24, Princeton, 1950, 51–72.

Bracken, J., and G. P. McCormick: "Selected Applications of Nonlinear Programming," John Wiley & Sons, Inc., New York, 1968.

Brondsted, A.: Conjugate Convex Functions in Topological Vector Spaces, *Matematisk-fysiske Meddelel ser udgivet af Det Kongelige Danske Videnskabernes Selskab*, **34**(2): 1–27 (1964).

Browder, F. E.: On the Unification of the Calculus of Variations and the Theory of Monotone Nonlinear Operators in Banach Spaces, *Proc. Nat. Acad. Sci. U.S.*, **56**: 419–425 (1966).

Buck, R. C.: "Advanced Calculus," McGraw-Hill Book Company, New York, 1965.

Canon, M., C. Cullum, and E. Polak: Constrained Minimization Problems in Finite-dimensional Spaces, *Society for Industrial and Applied Mathematics Journal on Control*, **4**: 528–547 (1966).

Carathéodory, C.: Über den variabilitätsberich der Koeffizienten von Potenzreihen, die gegebene Werte nicht annehmen, *Mathematische Annalen* **64**: 95–115 (1907).

Charnes, A., and W. W. Cooper: "Management Models and Industrial Applications of Linear Programming," vols. I, II, John Wiley & Sons, Inc., New York, 1961.

Cottle, R. W.: Symmetric Dual Quadratic Programs, *Quarterly of Applied Mathematics* **21**: 237–243 (1963).

Courant, R.: "Differential and Integral Calculus," vol. II, 2d ed., rev., Interscience Publishers, New York, 1947.

Courant, R., and D. Hilbert: "Methods of Mathematical Physics," pp. 231–242, Interscience Publishers, New York, 1953.

Dantzig, G. B.: "Linear Programming and Extensions," Princeton University Press, Princeton, N.J., 1963.

Dantzig, G. B., E. Eisenberg, and R. W. Cottle: Symmetric Dual Nonlinear Programs, *Pacific Journal of Mathematics*, **15**: 809–812 (1965).

Dennis, J. B.: "Mathematical Programming and Electrical Networks," John Wiley & Sons, Inc., New York, 1959.

Dieter, U.: Dualitat bei konvexen Optimeirungs—(Programmierungs—)Aufgaben, *Unternehmensforschung* **9**: 91–111 (1965a).

Dieter, U.: Dual External Problems in Locally Convex Linear Spaces, *Proceedings of the Colloquim on Convexity*, Copenhagen, 52–57 (1965b).

Dieter, U.: Optimierungsaufgaben in topologischen Vektorräumen I: Dualitätstheorie, *Zeitschift für Wahrscheinlichkeitstheorie und Verwandte Gebiete*, **5**: 89–117 (1966).

Dorn, W. S.: Duality in Quadratic Programming, *Quarterly of Applied Mathematics*, **18**: 155–162 (1960).

Dorn, W. S.: Self-dual Quadratic Programs, *Society for Industrial and Applied Mathematics Journal on Applied Mathematics*, **9**: 51–54 (1961).

Duffin, R. J.: Infinite Programs, in [Kuhn-Tucker 56], pp. 157–170.

Fan, K., I. Glicksburg, and A. J. Hoffman: Systems of Inequalities Involving Convex Functions, *American Mathematical Society Proceedings*, **8**: 617–622 (1957).

Farkas, J.: Über die Theorie der einfachen Ungleichungen, *Journal für die Reine und Angewandte Mathematik*, **124**: 1–24 (1902).

Fenchel, W.: "Convex Cones, Sets and Functions," Lecture notes, Princeton University, 1953, Armed Services Technical Information Agency, AD Number 22695.

Fiacco, A. V.: Second Order Sufficient Conditions for Weak and Strict Constrained Minima, *Society for Industrial and Applied Mathematics Journal on Applied Mathematics*, **16**: 105–108 (1968).

Fiacco, A. V., and G. P. McCormick: "Nonlinear Programming: Sequential Unconstrained Minimization Techniques," John Wiley & Sons, Inc., New York, 1968.

Fleming, W. H.: "Functions of Several Variables," McGraw-Hill Book Company, New York, 1965.

Friedrichs, K. O.: Ein Verfahren der Variationsrechnung des Minimum eines Integral als das Maximum eines Anderen Ausdruckes Dazutstellen, *Nachrichten von der Gesellschaft der Wissenschaften zu Göttingen Mathematische—Physikalische Klasse*, 13–20 (1929).

Gale, D.: "The Theory of Linear Economic Models," McGraw-Hill Book Company, New York, 1960.

Gale, D., H. W. Kuhn, and A. W. Tucker: Linear Programming and the Theory of Games, in [Koopmans 51], pp. 317–329.

Gass, S.: "Linear Programming," 2d ed., McGraw-Hill Book Company, New York, 1964.

Gol'stein, E. G.: Dual Problems of Convex and Fractionally-convex Programming in Functional Spaces, *Soviet Mathematics–Doklady* (English translation), **8**: 212–216 (1967).

Gordan, P.: Über die Auflösungen linearer Gleichungen mit reelen coefficienten, *Mathematische Annalen*, **6**: 23–28 (1873).

Graves, R. L., and P. Wolfe (ed.): "Recent Advances in Mathematical Programming," McGraw-Hill Book Company, New York, 1963.

Hadley, G. F.: "Linear Programming," Addison-Wesley Publishing Company, Inc., Mass., 1962.

Hadley, G. F.: "Nonlinear and Dynamic Programming," Addison-Wesley Publishing Company, Inc., Reading, Mass., 1964.

Halkin, H.: A Maximum Principle of the Pontryagin Type for Systems Described by Nonlinear Difference Equations, *Society for Industrial and Applied Mathematics Journal on Control*, **4**: 90–111 (1966).

Halkin, H., and L. W. Neustadt: General Necessary Conditions for Optimization Problems, *Proc. Nat. Acad. Sci. U.S.*, **56**: 1066–1071 (1966).

Halmos, P. R.: "Finite-dimensional Vector Spaces," D. Van Nostrand Company, Inc., Princeton, N.J., 1958.

Hamilton, N. T., and J. Landin: "Set Theory: The Structure of Arithmetic," Allyn and Bacon, Inc., Boston, 1961.

Hanson, M. A.: A Duality Theorem in Nonlinear Programming with Nonlinear Constraints, *Australian Journal of Statistics* **3**: 64–71 (1961).

Hanson, M. A.: Bounds for Functionally Convex Optimal Control Problems, *Journal of Mathematical Analysis and Applications*, **8**: 84–89 (1964).

[Hanson, M. A.: Duality for a Class of Infinite Programming Problems, *Society for Industrial and Applied Mathematics Journal on Applied Mathematics*, **16**: [318–323 (1968).

Hanson, M. A., and B. Mond: Quadratic Programming in Complex Space, *Journal of Mathematical Analysis and Applications*, **23**: 507–514 (1967).

Hestenes, M. R.: "Calculus of Variations and Optimal Control Theory," John Wiley & Sons, Inc., New York, 1966.

Hu, T. C.: "Integer Programming and Network Flows," Addison-Wesley Publishing Company, Inc., Reading, Mass., 1969.

Huard, P.: Dual Programs, *IBM J. Res. Develop.*, **6**: 137–139 (1962).

Hurwicz, L.: Programming in Linear Spaces, in [Arrow et al. 58], pp. 38–102.

Jacob, J.-P., and P. Rossi: "General Duality in Mathematical Programming," IBM Research Report, 1969.

Jensen, J. L. W. V.: Sur les fonctions convexes et les inégalités entre les valeurs moyennes, *Acta Mathematica*, **30**: 175–193 (1906).

John, F.: Extremum Problems with Inequalities as Subsidiary Conditions, in K. O. Friedrichs, O. E. Neugebauer, and J. J. Stoker, (eds.), "Studies and Essays: Courant Anniversary Volume," pp. 187–204, Interscience Publishers, New York, 1948.

Karamardian, S.: "Duality in Mathematical Programming," Operations Research Center, University of California, Berkeley, Report Number 66-2, 1966; *Journal of Mathematical Analysis and Applications*, **20**: 344–358 (1967).

Karlin, S.: "Mathematical Methods and Theory in Games, Programming, and Economics," vols. I, II, Addison-Wesley Publishing Company, Inc., Reading, Mass., 1959.

Koopman, T. J. (ed.): "Activity Analysis of Production and Allocation," John Wiley & Sons, Inc., New York, 1951.

Kowalik, J., and M. R. Osborne: "Methods for Unconstrained Optimization Problems," American Elsevier Publishing Company, Inc., New York, 1968.

Kretschmar, K. S.: Programmes in Paired Spaces, *Canadian Journal of Mathematics*, **13**: 221–238 (1961).

Kuhn, H. W., and A. W. Tucker: Nonlinear Programming, in J. Neyman (ed.), "Proceedings of the Second Berkeley Symposium on Mathematical Statistics and Probability," pp. 481–492, University of California Press, Berkeley, Calif., 1951.

Kuhn, H. W., and A. W. Tucker: "Linear Inequalities and Related Systems," Annals of Mathematics Studies Number 38, Princeton University Press, Princeton, N.J., 1956.

Künzi, H. P., W. Krelle, and W. Oettli: "Nonlinear Programming," Blaisdell Publishing Company, Waltham, Mass., 1966.

Larsen, A., and E. Polak: Some Sufficient Conditions for Continuous-linear-programming Problems," *International Journal of Engineering Science*, **4**: 583–604 (1966).

Lax, P. D.: Reciprocal Extremal Problems in Function Theory, *Communications on Pure and Applied Mathematics*, **8**: 437–453 (1955).

Levinson, N.: Linear Programming in Complex Space, *Journal of Mathematical Analysis and Applications*, **14**: 44–62 (1966).

Levitin, E. S., and B. T. Polyak: Constrained Minimization Methods, *USSR Computational Mathematics and Mathematical Physics* (English translation), **6**(5):1–50 (1966).

McCormick, G. P.: Second Order Conditions for Constrained Minima, *Society for Industrial and Applied Mathematics Journal on Applied Mathematics*, **15**: 641–652 (1967).

Mangasarian, O. L.: Duality in Nonlinear Programming, *Quarterly of Applied Mathematics*, **20**: 300–302 (1962).

Mangasarian, O. L.: Pseudo-convex Functions, *Society for Industrial and Applied Mathematics Journal on Control*, **3**: 281–290 (1965).

Mangasarian, O. L., and S. Fromovitz: A Maximum Principle in Mathematical Programming, in A. V. Balakrishnan and L. W. Neustadt (eds.), "Mathematical Theory of Control," pp. 85–95, Academic Press Inc., New York, 1967.

Mangasarian, O. L., and S. Fromovitz: The Fritz John Necessary Optimality Conditions in the Presence of Equality and Inequality Constraints, *J. Math. Analysis and Applications*, **17**:37–47 (1967a).

Mangasarian, O. L., and J. Ponstein: Minmax and Duality in Nonlinear Programming, *Journal of Mathematical Analysis and Applications*, **11**: 504–518, (1965).

Martos, B.: The Direct Power of Adjacent Vertex Programming Methods, *Management Science*, **12**: 241–252 (1965).

Minty, G. J.: On the Monotonicity of the Gradient of a Convex Function, *Pacific Journal of Mathematics*, **14**: 243–247 (1964).

Mond, B.: A Symmetric Dual Theorem for Nonlinear Programs, *Quarterly of Applied Mathematics*, **23**: 265–269 (1965).

Mond, B., and R. W. Cottle: Self-duality in Mathematical Programming, *Society for Industrial and Applied Mathematics Journal on Applied Mathematics*, **14**: 420–423 (1966).

Mond, B., and M. A. Hanson: Duality for Variational Problems, *Journal of Mathematical Analysis and Applications*, **18**: 355–364 (1967).

Moreau, J.-J.: Proximité et dualité dans un espace Hilbertien, *Bulletin de la Société Mathématique de France*, **93**: 273–299 (1965).

Motzkin, T. S.: "Beiträge zur Theorie der Linearen Ungleichungen," Inaugural Dissertation, Basel, Jerusalem, 1936.

Nikaidô, H.: On von Neumann's Minimax Theorem, *Pacific Journal of Mathematics*, **4**: 65–72 (1954).

Opial, Z.: "Nonexpansive and Monotone Mappings in Banach Spaces," Lecture Notes 67-1, Division of Applied Mathematics, Brown University, Providence, Rhode Island, 1967.

Ponstein, J.: Seven Types of Convexity, *Society for Industrial and Applied Mathematics Review*, **9**: 115–119 (1967).

Pontryagin, L. S., V. G. Boltyanskii, R. V. Gamkrelidze, and E. F. Mishchenko: "The Mathematical Theory of Optimal Processes," John Wiley & Sons, Inc., New York, 1962.

Rissanen, J.: On Duality Without Convexity, *Journal of Mathematical Analysis and Applications*, **18**: 269–275 (1967).

Ritter, K.: Duality for Nonlinear Programming in a Banach Space, *Society for Industrial and Applied Mathematics Journal on Applied Mathematics*, **15**: 294–302 (1967).

Rockafellar, R. T.: "Convex Functions and Dual Extremum Problems," Doctoral dissertation, Harvard University, 1963.

Rockafellar, R. T.: Duality Theorems for Convex Functions, *Bulletin of the American Mathematical Society*, **70**: 189–192 (1964).

Rockafellar, R. T.: Conjugates and Legendre Transforms of Convex Functions, *Canadian Journal of Mathematics*, **19**: 200–205 (1967a).

Rockafellar, R. T.: Convex Programming and Systems of Elementary Monotonic Relations, *Journal of Mathematical Analysis and Applications*, **19**: 543–564 (1967b).

Rockafellar, R. T.: Duality and Stability in Extremum Problems Involving Convex Functions, *Pacific Journal of Mathematics*, **21**: 167–187 (1967c).

Rockafellar, R. T.: "Convex Analysis," Princeton University Press, Princeton, N.J., 1969.

Rosen, J. B.: The Gradient Projection Method for Nonlinear Programming, *Society for Industrial and Applied Mathematics Journal on Applied Mathematics*, **8**: 181–217 (1960) and **9**: 514–553 (1961).

Rubinstein, G. S.: Dual Extremum Problems, *Soviet Mathematics–Doklady* (English translation), **4**: 1309–1312 (1963).

Rudin, W.: "Principles of Mathematical Analysis," 2d ed., McGraw-Hill Book Company, New York, 1964.

Saaty, T. L., and J. Bram: "Nonlinear Mathematics," McGraw-Hill Book Company, New York, 1964.

Simmonard, M.: "Linear Programming," Prentice-Hall, Inc., Englewood Cliffs, N.J., 1966.

Simmons, G. F.: "Introduction to Topology and Modern Analysis," McGraw-Hill Book Company, New York, 1963.

Slater, M.: "Lagrange Multipliers Revisited: A Contribution to Nonlinear Programming," Cowles Commission Discussion Paper, Mathematics 403, November, 1950.

Slater, M. L.: A Note on Motzkin's Transposition Theorem, *Econometrica*, **19**: 185–186 (1951).

Stiemke, E.: Über positive Lösungen homogener linearer Gleichungen, *Mathematische Annalen*, **76**: 340–342 (1915).

Stoer, J.: Duality in Nonlinear Programming and the Minimax Theorem, *Numerische Mathematik*, **5**: 371–379 (1963).

Stoer, J.: Über einen Dualitätssatz der nichtlinearen Programmierung, *Numerische Mathematik*, **6**: 55–58 (1964).

Trefftz, E.: "Ein Gegenstück zum Ritzschen Verfahren," Verhandlungen des Zweiten Internationalen Kongresses für Technische Mechanik, p. 131, Zürich, 1927.

Trefftz, E.: Konvergenz und Fehlerschätzung beim Ritzschen Verfahren, *Mathematische Annalen*, **100**: 503–521 (1928).

Tucker, A. W.: Dual Systems of Homogeneous Linear Relations, in [Kuhn-Tucker 56], pp. 3–18.

Tuy, Hoàng: Sur les inégalités linéaires, *Colloquium Mathematicum*, **13**: 107–123 (1964).

Tyndall, W. F.: A Duality Theorem for a Class of Continuous Linear Programming Problems, *Society for Industrial and Applied Mathematics Journal on Applied Mathematics*, **13**: 644–666 (1965).

Tyndall, W. F.: An Extended Duality Theorem for Continuous Linear Programming Problems, *Notices of the American Mathematical Society,* **14:** 152–153 (1967).

Uzawa, H.: The Kuhn-Tucker Theorem in Concave Programming, in [Arrow et al. 58], pp. 32–37.

Vajda, S.: "Mathematical Programming," Addison-Wesley Publishing Company, Inc., Reading, Mass., 1961.

Valentine, F. A.: "Convex Sets," McGraw-Hill Book Company, New York, 1964.

Van Slyke, R. M., and R. J. B. Wets: "A Duality Theory for Abstract Mathematical Programs with Applications to Optimal Control Theory," Mathematical Note Number 538, Mathematics Research Laboratory, Boeing Scientific Research Laboratories, October, 1967.

Varaiya, P. P.: Nonlinear Programming in Banach Space, *Society for Industrial and Applied Mathematics Journal on Applied Mathematics,* **15:** 284–293 (1967).

von Neumann, J.: Zur Theorie der Gesellshaftspiele, *Mathematische Annalen,* **100:** 295–320 (1928).

Whinston, A.: Conjugate Functions and Dual Programs, *Naval Research Logistics Quarterly,* **12:** 315–322 (1965).

Whinston, A.: Some Applications of the Conjugate Function Theory to Duality, in [Abadie 67], pp. 75–96.

Wolfe, P.: A Duality Theorem for Nonlinear Programming, *Quarterly of Applied Mathematics,* **19:** 239–244 (1961).

Zangwill, W. I.: "Nonlinear Programming: A Unified Approach," Prentice-Hall, Inc., Englewood Cliffs, N.J., 1969.

Zarantonello, E. H.: "Solving Functional Equations by Contractive Averaging," Mathematics Research Center, University of Wisconsin, Technical Summary Report Number 160, 1960.

Zoutendijk, G.: "Methods of Feasible Directions," Elsevier Publishing Company, Amsterdam, 1960.

Zukhovitskiy, S. I., and L. I. Avdeyeva: "Linear and Convex Programming," W. B. Saunders Company, Philadelphia, 1966.

Indexes

Name Index

Subject Index

This book was set in Modern by The Maple Press Company, and printed on permanent paper and bound by The Maple Press Company. The designer was Marsha Cohen; the drawings were done by Engineering-Drafting Company. The editors were B. Dandison, Jr. and Maureen McMahon. Peter D. Guilmette supervised the production.